ISBN 978-0-282-77661-9
PIBN 10226949

1 MONTH OF
FREE
READING

at

www.ForgottenBooks.com

---◊---

By purchasing this book you are eligible for one month membership to ForgottenBooks.com, giving you unlimited access to our entire collection of over 700,000 titles via our web site and mobile apps.

To claim your free month visit:
www.forgottenbooks.com/free226949

English
Français
Deutsche
Italiano
Español
Português

www.forgottenbooks.com

Mythology Photography **Fiction**
Fishing Christianity **Art** Cooking
Essays Buddhism Freemasonry
Medicine **Biology** Music **Ancient
Egypt** Evolution Carpentry Physics
Dance Geology **Mathematics** Fitness
Shakespeare **Folklore** Yoga Marketing
Confidence Immortality Biographies
Poetry **Psychology** Witchcraft
Electronics Chemistry History **Law**
Accounting **Philosophy** Anthropology
Alchemy Drama Quantum Mechanics
Atheism Sexual Health **Ancient History**
Entrepreneurship Languages Sport
Paleontology Needlework Islam
Metaphysics Investment Archaeology
Parenting Statistics Criminology
Motivational

APPLETONS'

ILLUSTRATED

HAND-BOOK OF AMERICAN TRAVEL.

THE EASTERN AND MIDDLE STATES, AND THE BRITISH PROVINCES.

NEW YORK:

APPLETON & CO., 443 & 445 BROADWAY.

LONDON: TRÜBNER & CO.

1861.

NOTICE.

No expense or labor will be spared to make the Hand-Book of Ame
ican Travel attractive, comprehensive, concise, thorough, and every wა
reliable.

The next Annual Edition will be published in May, 1861, and aı
information in regard to errors and omissions, which those who use tl
work may detect, or any facts of interest and value—particularly
respect to new routes and accommodations—will be gratefully receiv
and considered. Such communications should be addressed to the Autho
care of the Publishers

☞ For CONTENTS, GENERAL INDEX, LIST OF MAPS, LIST OF ILLU
TRATIONS, etc., see the end of the Volume.

The Populations of Cities and Towns mentioned in this work are tho
given in the last National Census—1850—except when otherwise statec

APPLETONS'
ILLUSTRATED RAILWAY AND STEAM NAVIGATION
GUIDE,

Containing Seventy Maps, and the latest Time Tables, Corrected to Date. Pu
lished Monthly, under the supervision of the Railroad Companies. One Volun
288 pages. Price, Twenty-five Cents. Subscription price, Three Dollars p
Annum.

APPLETONS'

ILLUSTRATED HAND-BOOK

OF

AMERICAN TRAVEL.

A FULL AND RELIABLE

GUIDE

BY RAILWAY, STEAMBOAT, AND STAGE,

TO THE CITIES, TOWNS, WATERFALLS, BATTLE-FIELDS, MOUNTAINS, RIVERS, LAKES, HUNT-
ING AND FISHING GROUNDS, WATERING PLACES, SUMMER RESORTS, AND ALL
SCENES AND OBJECTS OF IMPORTANCE AND INTEREST IN

THE UNITED STATES AND THE BRITISH PROVINCES.

BY

T. ADDISON RICHARDS.

WITH CAREFUL MAPS OF ALL PARTS OF THE COUNTRY, AND PICTURES OF FAMOUS PLACES
AND SCENES, FROM ORIGINAL DRAWINGS BY THE AUTHOR AND OTHER ARTISTS.

ENGRAVED AND ELECTROTYPED BY WHITNEY AND JOCELYN.

"When thou haply seest
Some rare, note-worthy object in thy travels,
Make me partaker of thy happiness."—*Shakspeare.*

.

NEW YORK:
D. APPLETON & CO., 443 & 445 BROADWAY.
LONDON: TRÜBNER & CO.
1861.

TO THE TRAVELLER:

SOME PARTING WORDS OF EXPLANATION AND ADVICE.

In a journey over so vast a country as the United States, occupying nearly half a Continent, and measuring its length and breadth by thousands, and its routes of travel by tens of thousands of miles, one may very readily be pardoned if he sometimes stumbles by the way. May *we* not beg the benefit of this consideration, if, in our present laborious *itinéraire*, we have occasionally chanced, despite all our watchfulness, to only half look at points of interest or to overlook them altogether; or if, amidst the intricate riticulation of the roads, we may have momently lost our way? We hope, however, that we have not been thus unlucky in any considerable degree, for we have made very honest effort to guide our traveller truly and surely ; to show him—hastily, to be sure, as needs must be, yet intelligently—the past and the present, the physique and the *morale*, of the great country through which we have led him ; its differing peoples and places, from the mountains to the prairies —from the cities and palaces of the East to the wildernesses and wigwams of the West.

Though we have thus done our best for the present, we hope to do still better hereafter, as we revise and extend our volume year after year, with the benefit of enlarged personal observation and of the good counsels of others : for we trust that those who follow our guidance will do us the kindness to advise us of any and all errors and omissions they may discover in our pages. To assist them in rendering us this generous service, we have placed some blank leaves for memoranda, at the end of our book.

In our list of illustrations, we have the good fortune to include valuable contributions from the portfolios of our gifted brother-artists, Mr. James H. Cafferty, Mr. F. E. Church, Mr. Jervis McEntee, Mr. A. D. Shattuck, Mr. S. S. David, Mr. Charles Lanman, Mr. T. A. Ayres, and others. In the literary department, we are much indebted to Mr. Ayres, for the careful account of California; to Mr. T. D. Lowther, for the very pleasant mention of St. Augustine; and to William Prescott Smith, for the account of the Baltimore and Ohio Railroad.

THE PLAN OF THIS BOOK.

We have thought it best to follow the familiar geographical order of the various divisions of the country, and thus to begin at Canada on the extreme north-east, and, continuing along the shores of the Atlantic and the Gulf of Mexico, end upon the Pacific, westward. With rare exceptions, we have, instead of selecting a particular route and seeing all it offers of attraction, jumped at once to our especial destination, and then intimated the way by which it is reached. Thus, if the traveller happens to be in New York or Boston, and desires to go to New Orleans, he will, by turning to "New Orleans," find the routes thither. The chief cities are taken as starting points for all other and lesser places in their neighborhood. It has not, of course, been possible to mention every village or town in the Union, in the narrow limits of a pocket volume, like this.

GENERAL REMARKS.

The foreign tourist will soon observe, to his satisfaction, (and the *citizen* might remember it oftener, with thanks to his stars,) the great convenience of the total absence in the United States, of all annoying demands for passports—of scowling fortifications and draw-bridges, of jealous gates, closed at a fixed hour of the evening and not to be reopened before another fixed hour of the morning; of custom-houses between the several States, and of all rummaging of baggage by gens d'armes for the octroi; and yet nevertheless, of as perfect a feeling of security, every where, as in the most vigilantly policed kingdoms of Europe.

He may or may not like the *table d'hôte* system of our hotels —the uniform fare and the unvarying price; that, excepting in the few metropolitan cities, where the habits of all nations obtain, we must submit to.

From the social equality every where and without exception, he will not suffer, however high his rank at home; and if it be not the highest, he will surely gain in consideration. To win attention and care, both the lofty and the lowly have, and have only, to dispense good will and kind manners as they pass along.

MONEY.

Gold and silver, it should be remembered, are always and every where current, while bank-notes, and especially of distant States, very often are not. Change, too, will save trouble; especially half-dollars, generally the fare of omnibuses and hacks, and invariably the price of meals. Twenty-five cent pieces, too, are useful, as fees for little services by the way. In travelling through the settled districts by railways and steamboats, and at the best hotels, the daily expenses should be estimated at not less than five or six dollars per day for each person.

BAGGAGE.

As little baggage as possible is always a good rule, though a very liberal supply is permitted on the railways and almost any quantity on the steamboats. On the stages, the prescribed limit of sixty or eighty pounds cannot be exceeded without extra charge.

The regular carriages of his hotel will convey the traveller securely and in season, to the railway station or the steamboat landing, where his first care must be to deposit his trunks in the keeping of the baggage-master, and receive a check for each one— corresponding marks will be attached to the baggage, and it will be delivered at the end of the route only to the holder of the checks. It is best to get baggage checked for the entire journey, or for the longest possible stage thereof, and thus save one's self the trouble of looking out for it more frequently than is necessary.

Before arriving at his destination, the traveller will, on the principal routes, receive a call from an express agent, to whom he may safely resign his check and his address, confident that his baggage will be duly delivered, and at the fixed tariff of twenty-five cents for each piece or trunk. On arriving at the end of his journey, he should put himself in one of the carriages marked as in the particular service of the hotel to which he is going. If he employs other vehicles, it will be well to learn the fare beforehand, particularly in the city of New York, where hackmen pay but little attention when they can help it, to the law in the case.

TICKETS.

Tickets on the railways should be purchased at the office before starting, otherwise a small additional charge will be made. If a long journey over various roads is intended, it is cheaper and more convenient to buy a through ticket to the end of the route, or for as long a distance as possible. On the steamboats, the tickets for passage and for meals will be purchased at leisure, after starting, at the captain's office.

HOTELS.

The hotels of the United States are famous all the world over, for their extent, convenience, comfort and elegance. They are often truly palatial in their sumptuousness, with means and appliances for the prompt gratification of every want and whim. The universal price of board, from one end of the country to the other, is $2 50 or $2 00 per day at the most fashionable, and indeed at all the principal houses. Private parlors and extra rooms involve an additional charge, according to their position. Wines are always extra and always dear enough.

WAITERS OR SERVANTS.

It is not the general custom in America, as in Europe, to fee waiters at the hotels, though it may very properly be done for especial personal service. It is often done by those who like hot dinners better than cold, or who may have a fancy for some rare dish when it unluckily happens to be " all out."

COSTUME.

At the watering places, the same resources of toilette are

needed as in the city *salon ;* but though you be thus provided, do not be unprovided with a travelling suit equal to rude usage. If the color be a gray or a brown, so much the better in the dust of railway and stage routes. Don a felt hat,—it does not crush itself or your head in car or carriage, or blow overboard on steamboats. Leave thin boots (this especially to the ladies) at home, and be well, and comfortably, and safely shod, in *stout calf skin*. It is a pity to be kept in doors by the fear of spoiling one's gaiters or wetting one's feet, when the meadows and hills and brooks are waiting to be explored. In mountain tramps, a generous sized flask, filled with most excellent brandy, may be swung over the shoulder with very picturesque effect.

Now that we have told our traveller *how* to go, it only remains to us, before starting, to add a word of suggestion as to

WHERE TO GO.

If you are in New York, with one or two or three, or more summer days to spare, run up to one or other of the many delightful places on the Hudson River,—to West Point, or Newburgh, for example; or to the Catskill Mountains; or run down to Rockaway, or Long Branch, or any of the many healthful and inviting resorts along the coast of Long Island and New Jersey.

If a week is at your command, go to Lake George, or to Trenton Falls, or Niagara; explore the varied route of the Erie Railway, or seek some one of the innumerable Springs of the State.

If a fortnight or a month can be spared, make a trip to Canada. See Montreal and the Ottawa River, then go to Quebec and the Saguenay, returning through Maine; or from Montreal go up the St. Lawrence to Toronto, and thence to the great Lakes; or spend a part or all of your time among the wonderful White Hills of New Hampshire.

If the whole summer is waiting to be disposed of, visit the countless watering-places in the mountain lands of Western Virginia; or see the landscape beauties of the Blue Ridge regions of the Carolinas and Georgia; or astonish yourself with a glimpse of the western cities and of the Mammoth Cave.

1*

In winter leisure, go to Charleston, and Savannah, and New Orleans, Florida and Cuba, and find the summer airs again which you have lost in higher latitudes.

There is no lack of inviting resorts for a day, or week, or month, or for ever. Look in this respect at our Skeleton Tours, and at the detailed descriptions and routes in the pages which follow. *Go somewhere, if you can,* all of you, and wherever and whenever you go, God speed you on your way and send you duly back wiser, and better, and healthier, and happier men and women.

AMERICAN FISHING AND FIELD SPORTS.

WE cannot well turn a thought to the boundless resources in forest and flood, which the great and varied territory of the United States presents for every description of field sport, without a sigh of regret that they should be so little esteemed and employed by the people. Nowhere are the wildernesses and the woods so populous with the noblest objects of the chase, the bison, the bear and the deer of every species, while the waters of the numberless inlets and estuaries of the great bays of our Atlantic coast, perhaps surpass those of any other country in the world in their immense supplies of all kinds of wild fowl.

Opportunity waits the will and pleasure alike of the hardiest and boldest forest adventurer, and of the daintest dilettante of the mossy brook-side. All may find abundant occupation for their differing tastes, and with no jealous or unreasonable legal let or hinderance to the full indulgence thereof, as in some other less free and favored lands.

The health, moral no less than physical, which is ever to be gathered in the exercise and pleasures of the chase, commends it, especially, to a people disposed, like ours, to over-work and over-toil; and if the scholar could be persuaded, sometimes, to close his wearying books, the merchant to leave his dull desk, and the artisan his unceasing toil for a generous indulgence in out-of-door relaxation, who can tell how much the enjoyment of life might be increased—how greatly life itself might be prolonged?

For the angler there is opportunity every where in the mountain brooks and woodland streams, and in the innumerable rivers and lakes of the land.

Sea trout and salmon, of the finest size and quality, abound in Maine, east of the mouth of the Kennebec. In Nova Scotia, New Brunswick and Canada, too, they are most plentiful. The salmon, especially, may be readily taken in the waters of the St. John's river, in the St. Lawrence, as far up as the "Thousand Islands," and down almost to the sea; in the Chaudiere, the St. Maurice, the Jacques Cartier and the Ottawa. From Lake Ontario, this fine fish makes its way through the Oswego river into Lake Cayuga, and it is said enters Seneca Lake also. In these last-mentioned waters they are usually taken with the net, and very rarely, if ever, by the fly.

Smaller trout of exquisite flavor are abundant in all the mountain streams of the Northern and the Middle States, as far southward as Virginia. They swarm, also, in Lake Superior, where the salmon, too, is to be found. Long Island is famous for its trout; and so, also, is the wilderness of northern New York—all that beautiful mountain and lake region lying west of Champlain. The brook or spotted trout here, are often taken up to two and a-half or three pounds, and sometimes up to six, though a two-pounder is usually esteemed a god-send, as the average size of this fish is much less in American than in European waters.

The black bass is plentiful in the lakes, and the pike, the pickerel, the maskalonge and the striped bass, besides many lesser kind.

The most abundant facilities for wild-fowl shooting may be found at all points of the long reach of the Atlantic coast—in the noble bays and the rivers—inlets from Maine to Florida. At some one or other period of the year, each species of wader, from the stately swan to the little sand-piper, take up their abode in every point of this region—now flocking to the land-locked bays and lagunes of Long Island and New Jersey; now about the estuaries of the great Chesapeake, and later still, in the Albemarle and the Pamlico Sound, off Carolina. See "Long Island" in this respect, and the "Chesapeake Bay" in regard to the famous canvas-back.

Even to catalogue the eligible spots for every kind of upland shooting in a country covered with forests and woods, and necessarily filled with birds and other game of all kinds, would demand a volume. The sportsman may choose for himself where to kill the prairie hen, any where from Texas up, through all the great western plains; the ruffed grouse, in New England where it is called partridge, and in Pennsylvania, where it is known as the pheasant; or the Canada grouse in upper New York and northward. The quail may be found abundantly from Massachusetts to Ohio and Kentucky, and even to Texas. The woodcock is plentiful in all the Eastern and Middle States, and in winter not less so in the South. Ducks, too, of many kinds, and teal and rail will every where repay the search of the hunter.

Of the nobler game of the forest-wilds, the rude bison, the lordly elk and moose, the deer, and even of the bear and the panther, not only the great Western wildernesses, but the mountain glens of the northern States, and the jungles of the South, offer ample supplies. The deer, more especially, is yet to be found in every part of the United States, and abundantly in the woods of Maine and the mountain region of all New England and of northern New York. See chapters on the Adirondack and the Saranac lake region. Hereabouts this beautiful animal is most often killed by one or other of the two questionable modes, called driving and still hunting; but south of the Potomac, in Virginia, the Carolinas, Mississippi and Louisiana, he is yet followed in the good old brave manly chase, with all the inspiriting and ennobling adjuncts of mounted cavalier, daring flood and fell at the call of the bugle note and the stirring halloo.

The bison and the elk must be followed to the Far West, beyond the Mississippi, whither they have been driven long since, before the inexorable course of empire. How much longer they may be found even there, who can tell, while Civilization is stalking, as it is, over the great deserts in its annihilating seven-league boots.

"Let the man who's disturbed by misfortune and care,
Away to the woodlands and valleys repair.
Tally-ho, etc.

Let him hear but the notes of the sweet swelling horn,
With the hounds in full cry, and his troubles are gone.
Tally-ho, etc."
OLD SONG

SKELETON TOURS

From New York to Various Parts of the United States and the Canadas.

WITH AN APPROXIMATE STATEMENT OF THE TIME REQUIRED TO TRAVEL
FROM PLACE TO PLACE, AND OF THE DURATION OF THE HALTS
TO BE MADE AT THE MOST REMARKABLE SPOTS.

*See Description of Routes, Hotels, Places and Scenes in the Following Pages.**

A TOUR OF SIX DAYS,
Visiting West Point, Newburgh, and the Catskills.

MONDAY. New York to West Point (52 miles), by morning steamboat up the Hudson
River, through the Highlands, or by an early train on the Hudson River Railway,
stopping at Garrison's, and crossing by steam-ferry to the West Point Hotel or to
Cozzens', just below. Arrive in three hours, by or before noon (page 128).

Visit the Military Academy, the ruins of Fort Putnam, Kosciusko's Garden,
Weir's Studio, etc.

TUESDAY. Morning steamboat or early train to Newburgh (9 miles, crossing ferries
included, one hour), stop at the Orange Hotel, on the Main street, or at the Powelton, an elegant summer establishment in the suburbs; visit Washington's
Head Quarters in the village. After dinner, take a carriage for "Idlewild," the
charming home of N. P. Willis, four miles down the river. Explore the grounds
and the beautiful mountain brook and glen. Visit "Cedarlawn," the residence
of the author Headley, on the way, a mile below Newburgh (pages 130, 131).

WEDNESDAY. Morning steamboat or by Railway from Fishkill, opposite Newburgh,
to Oakhill, opposite Catskill; 51 miles, 2 hours, *besides ferries*. From Catskill
village, in good coaches, 12 miles, through a most picturesque hill and valley region, to the Mountain House (page 144).

THURSDAY. Look out for the grand spectacle, from this point, of the sun-rise. After
breakfast walk to the North Mountain, overlooking the Hotel and the two lakes;
next, join the usual morning party in the two-mile ride to the High Falls: back
to dinner.

FRIDAY. Ride from the Mountain House through the great Kauterskill Clove, westward to the village of Palenville, returning by valley and mountain road eastward; or explore the ravines and cascades of the Clove, better *on foot*—a good
day's tramp.

SATURDAY. Return to New York, via Catskill Village and the Hudson River.

*** If more time is at command, devote a day to a visit to High Peak, another
to the Stony Clove, and another to the Plauterkill Clove and Creek.

---

* For Rail'way Time-Tables consult Appletons' Monthly Railway and Steam Navigation
Guide.

A TOUR OF SIX DAYS,

Visiting Albany and Troy (via the Hudson River), Saratoga Springs, Lake George, Fort Ticonderoga, and Whitehall, on Lake Champlain.

MONDAY. From New York by morning boat or cars, via Hudson River, 146 miles, 5 or 6 hours, to Albany (see Albany and Troy, pp. 134 and 135.)

TUESDAY. Railway, time about two hours, from Albany, through the city of Troy, to Saratoga Springs. Stop at the United States Hotel, or at the Union Hall (p.149).

WEDNESDAY. To Lake George by Railway 15 miles, to Moreau Station, and thence by plank road, an hour or two, via Glen's Falls to Caldwell. Stop over night at the Lake House, or at the Fort William Henry Hotel, close by (page 151).

THURSDAY. Spend the day on the Lake boating and fishing, or sketching.

FRIDAY. Make the voyage of the lake in the favorite little steamer, the "Minne-haha," a few hours, sail to the village of Ticonderoga, at the foot or north end of the lake; thence 3 or 4 miles by coach to the ruins of Fort Ticonderoga.

SATURDAY. Return home by the Lake Champlain steamers to Whitehall (page 154), and thence by Railway via Troy and the Hudson.

*** Same tour (except on the Hudson), within the same time from Boston, taking the Western Railway, thence (*Monday*) 200 miles to Albany.

A TOUR OF SIX DAYS,

Visiting Trenton and Niagara Falls, via the Central Railway, and Returning by the New York and Erie Road.

MONDAY. From New York to Trenton Falls, via Hudson River, 146 miles to Albany, Central Railway, 95 miles to Utica, thence 15 miles to the Falls (page 156).

TUESDAY. Explore the Falls.

WEDNESDAY. Return to Utica and resume journey on the Central Road, via Syracuse and Rochester (Falls of the Genesee) to Niagara (page 161).

THURSDAY. At Niagara.

FRIDAY. To Buffalo, and thence by the Erie Railway, passing the night at Binghamton.

SATURDAY. Erie Road from Binghamton to New York.

*** If more time is at command, remain over Sunday at Niagara, and follow the picturesque route of the Erie Road more leisurely, seeing the cascades and ravines of the Genesee, and the great Railway Bridge at Portage, 61 miles from Buffalo. Elmira, 273 miles from New York; Owego, 236 miles; Great Bend, 200 miles; and P rt Jervis, 88 miles, are pleasant stopping places on the way.

TOUR OF A WEEK,

Visiting Philadelphia, Baltimore, and Washington City.

MONDAY. From New York by morning line to Philadelphia on the New Jersey Railway, 87 miles, or by the Camden and Amboy route (pages 174 and 176). Arrive in the early afternoon.

TUESDAY. At Philadelphia (see page 182).

WEDNESDAY. Morning train to Baltimore (page 179), 97 miles; arrive in the early afternoon. For Baltimore see page 199.

THURSDAY. Spend the day in Baltimore, and take the evening train (page 214), 40 miles, 2 hours, to Washington.

FRIDAY. At Washington (page 215).

SATURDAY. Return to New York by Baltimore and Philadelphia, 224 miles.

*** If more time can be spared, remain in Washington Saturday and Sunday, visiting Mount Vernon (page 218), Georgetown, Alexandria, etc. Return on Monday to Philadelphia, and next day leisurely to New York.

TOUR OF A WEEK,

Visiting the Valley of Wyoming and the Delaware Water Gap.

MONDAY. From New York by the Erie Railway, 200 miles, to Great Bend.

TUESDAY. By the Delaware, Lackawanna and Western Railway to Scranton, an interesting place; thence to Wilkesbarre, on the Susquehanna, and in the Valley of Wyoming (page 194).

WEDNESDAY. Explore the valley, visiting Prospect Rock, 3 miles from the village and Nanticoke, in the beautiful passage of the Susquehanna, at the Southern extremity of Wyoming.

THURSDAY. Returning, take the cars via Mauch Chunk, in the coal region, to Easton and the Water Gap.

FRIDAY. At the Water Gap (see page 197.)

SATURDAY. Reach home by the Delaware, Lackawanna, and Western Railway, and other routes across New Jersey.

*** With more time, it would be agreeable to spend a day at Scranton, two or three in and below the Valley of Wyoming; to stop at Mauch Chunck, and see the coal mines and the bold landscape of the Lehigh River.

TOUR OF TWO WEEKS,

Visiting the White Mountains and the Lake Scenery of New Hampshire, via Boston.

FIRST WEEK.

MONDAY. From New York to Boston (page 61), journey occupying the night, by the Stonington and Providence, the Fall River or the Norwich routes (morning and evening), or by the Boston Express; or the Shore Line via New London.]

TUESDAY. Boston (page 65.)

WEDNESDAY. Boston to Centre Harbor, on Lake Winnipiseogee (page 88); arrive at dinner time; spend the afternoon on the lake or lake shores.

THURSDAY. Visit Red Hill (on horseback), a few miles distant, and overlook the beautiful lake region (page 91).

FRIDAY. Proceed, by the White Mountain stages, to North Conway, one of the most charming valleys in the world; stop over night at Thompson's (page 91).

SATURDAY. Continue journey by stage 24 miles to Crawford House, in the Great White Mountain Notch—traversing the valleys of Conway, Bartlett, etc.

SUNDAY. Crawford House (page 96).

SECOND WEEK.

MONDAY. Ascend Mount Washington (page 96).

TUESDAY. Visit the Silver Cascade and other scenes in the neighborhood of the Notch.

WEDNESDAY. Continue journey, by stage, 27 miles (page 99), to the Profile House, in the Franconia group of the White Hills, following the course of the Ammonoosuc.

THURSDAY. Profile House; visit Echo Lake and Profile Lake, and see the Old Man of the Mountain, Eagle Cliff, Cannon Mountain, and other sights of the vicinage.

FRIDAY. Ride 5 miles, from the Profile to the Flume House; visit the Flume and its neighboring marvels.

SATURDAY. Returning, take stage to Littleton, 12 miles; thence by Railway 20 miles to Wells' River; thence through the valley of the Connecticut to Bellows Falls, Brattleboro, or Northampton.

SUNDAY. At Bellows Falls, Brattleboro or Northampton.

THIRD MONDAY. Home, from Northampton, by Springfield, Hartford, and New Haven, or (from Bellows Falls) by Albany and the Hudson.

A TOUR OF TWO WEEKS,

From New York to the White Mountains, via Boston and Portland, Returning by the Connecticut Valley Routes.

FIRST WEEK.
MONDAY. New York to Boston (see routes, page 61).
TUESDAY. At Boston (page 65).
WEDNESDAY. Boston to Portland, Maine. (see routes, page 58).
THURSDAY. At Portland (page 58).
FRIDAY. From Portland by the Grand Trunk Railway, 91 miles to Gorham, N. H., White Mountain Station (page 94); continue journey in coaches 8 miles to the Glen House (page 94).
SATURDAY. Journey by stage from the Glen House, 34 miles, to the Crawford House, White Mountain Notch.
SUNDAY. At the Crawford House.

SECOND WEEK.
Explore the White Mountains and return home, as in preceding Tour.

A TOUR OF TWO WEEKS,

Visiting the New England Cities, New Haven, Hartford, Springfield, Boston, Providence, and Newport.

FIRST WEEK.
MONDAY. From New York to New Haven, Ct., 76 miles, by the New Haven Railway; visit Yale College, the Trumbull Gallery, etc. (page 62).
TUESDAY. Continue journey, 36 miles, to Hartford, Ct. (page 62).
WEDNESDAY. To Springfield, Mass., 26 miles (page 63); visit the United States Armory, etc.
THURSDAY. To Boston, 98 miles (page 65).
FRIDAY. At Boston (page 65).
SATURDAY. At Boston.
SUNDAY. At Boston.

SECOND WEEK.
MONDAY. Morning train from Boston, 43 miles, to Providence (page 83); see the Library of Brown University and the Atheneum; visit the Seekonk River and "What Cheer Rock," on the edge of the city, the village and Falls of Pawtucket, near by, etc.
TUESDAY. At Providence; take a sail down the Narragansett Bay and back, in one of the numerous excursion steamers (page 84).
WEDNESDAY. Take the steamboat down the Narragansett Bay, from Providence to Newport; a charming voyage of some two hours.
THURSDAY. At Newport (page 86)
FRIDAY. At Newport.
SATURDAY. At Newport.
SUNDAY. At Newport.
MONDAY. Home, by Fall River Steamers direct.

TOUR OF TWO WEEKS,

From New York up the Valley of the Housatonic to Great Barrington, Stockbridge, etc., in Berkshire, Mass.; to Lebanon Springs and Shaker Village, N. Y. Returning via the Hudson River and West Point.

FIRST WEEK.

MONDAY. From New York, via New Haven Railway (page 62), to Bridgeport, Ct.; thence, without stopping, by the Housatonic Railway (page 80) up the valley and river of the Housatonic to Great Barrington, in Berkshire, Mass.

TUESDAY. At Great Barrington (page 81).

WEDNESDAY. From Great Barrington, Railway 26 miles to Old Stockbridge.

THURSDAY. At Old Stockbridge (page 81).

FRIDAY. Lebanon Springs.

SATURDAY. Lebanon Springs (page 171).

SUNDAY. Lebanon Springs. Visit Shaker village, near by (page 171).

SECOND WEEK.

MONDAY. Visit Pittsfield, Williamstown, Lenox, Adams, etc.

TUESDAY. Visit Pittsfield, Williamstown, Lenox, Adams, etc.

WEDNESDAY. Visit Pittsfield, Williamstown, Lenox, Adams, etc.

THURSDAY. Return via Western Railway to Albany, or by the Hudson and Boston Road to Hudson, and thence down the Hudson River to West Point.

FRIDAY. At West Point (page 128).

SATURDAY. Back in New York.

TOUR OF TWO WEEKS,

Visiting the Valley of the Connecticut.

MONDAY. By Railway from New York via New Haven and Hartford, Ct., to Springfield, Mass., 138 miles; dine, visit the U. S. Armory, etc. (pages 61 to 63).

TUESDAY. To Northampton, 17 miles, by Railway, near the banks of the Connecticut.

WEDNESDAY. At Northampton (page 73), visiting Mount Holyoke, and other scenes of great interest in the immediate neighborhood.

THURSDAY. Continue on the Railways up the valley and river 19 miles to Greenfield, Mass.; walk in the evening to the high ridge called Poet's Seat, finely overlooking all the country round.

FRIDAY. Resume the journey (by Railway always), up the valley, 24 miles further, to Brattleboro, in Vermont. This is one of the most agreeable resting places on the route; one of the most attractive in scenery, society, hotel comforts, etc. (see page 78).

SATURDAY. Visit the grounds of the Insane Asylum, West River, the Cemetery, and other charming localities in the vicinage of Brattleboro.

SUNDAY. Still at Brattleboro; a pleasant place for a Sunday halt, all travel being suspended on that day hereabouts.

MONDAY. Resume journey 24 miles further up the river to Bellows Falls (page 79). At this point the traveller may turn back if he pleases by railway via Rutland, Vt., Whitehall, on Lake Champlain, Saratoga Springs, Albany or Troy, and the Hudson River; going on Tuesday to Saratoga, and on Wednesday to New York; or he may continue on with us yet further up the valley of the Connecticut.

TUESDAY. From Bellows Falls 26 miles to Windsor, Vt., a very quiet, picturesque, and pleasant place (page 79).

WEDNESDAY. Ascend Mount Ascutney, near Windsor.

THURSDAY. From Windsor (returning) by the Vermont Central Road, through the charming valley of the Winooski to Burlington, on Lake Champlain (page 101).

FRIDAY. Cross the Lake from Burlington to Port Kent, and visit the bold ravine called the Walled Banks of the Ausable.

SATURDAY. Home by Whitehall, Troy, Albany, and the Hudson.

₄ At Windsor (Second Tuesday of this tour), the traveller being on one of the most agreeable routes thence, may continue his journey eastward to the White Mountain Region.

TOUR OF THREE WEEKS,

Visiting the Hudson River, Saratoga Springs, Lake George, Lake Champlain, Montreal, Quebec, and the Saguenay River, the St. Lawrence River, Niagara Falls, and the Scenery of the Erie Railway.

FIRST WEEK.

MONDAY. From New York to Albany, by steamboat or railway (Hudson River), thence by railway to Saratoga (pages 119 and 149).

TUESDAY. Saratoga Springs (page 149).

WEDNESDAY. To Caldwell on Lake George (page 151).

THURSDAY. Down Lake George to Fort Ticonderoga on Lake Champlain (page 150 to 153).

FRIDAY. Steamer on Lake Champlain (a pleasant voyage) and onward by railway to Montreal.

SATURDAY. Montreal (page 32).

SUNDAY. Montreal.

SECOND WEEK.

MONDAY. Railway or St. Lawrence River to Quebec (page 35).

TUESDAY. Quebec (page 35).

WEDNESDAY. Down the St. Lawrence to the mouth of the Saguenay (page 29).

THURSDAY. Voyage up the Saguenay.

FRIDAY. Back to Quebec.

SATURDAY. Grand Trunk Railway to Montreal.

SUNDAY. Montreal.

THIRD WEEK.

MONDAY. Up the St. Lawrence to Kingston (page 41).

TUESDAY. Grand Trunk Railway via Toronto to Hamilton; thence, by the Great Western Road to Suspension Bridge, Niagara.

WEDNESDAY. Niagara Falls (page 161).

THURSDAY. Niagara.

FRIDAY. Erie Railway (returning) to Owego or Binghamton, or to Utica on the Central Route.

SATURDAY. Home,

₄ Omit the détour from Montreal to Quebec, and back, and make this tour within two weeks instead of three.

HUNTING TOUR OF THREE WEEKS,

To the Saranac Lakes, in the Wilderness of Northern New York.

FIRST WEEK.

MONDAY. From New York to Port Kent, opposite Burlington, on Lake Champlain, via Hudson River, Saratoga Springs, and Whitehall (page 153). From Port Kent, by omnibus or stage five miles back, to Keeseville. Stop at the Ausable House.

TUESDAY. Visit the remarkable ravines and cascades near Keeseville, called the Walled Banks of the Ausable (page 155).

WEDNESDAY. Take the tri-weekly mail wagon or private carriage, for the banks of the Lower Saranac Lake, stopping at Baker's, a mile distant, or at Martin's on the shore (page 168).

THURSDAY. Secure the services of a guide and hunter, with his boat, dogs, tent, and all necessary equipments and provisions for camp life, all the journey hence being by water (page 169).

FRIDAY. On the Lower Saranac, crossing the "carrying place" in the afternoon to the Middle Saranac, on the shore of which camp for the night, after a supper of trout, readily taken, with venison, perchance, to boot.

SATURDAY and SUNDAY. Camp on the Upper Saranac, one of the most beautiful of these lakes, and a fine hunting and fishing ground.

SECOND WEEK.

MONDAY and TUESDAY. Visit the St. Regis Lake.

WEDNESDAY. Return to the Middle Saranac (or Round Lake), make a short portage to the Stony Creek Pond ; and thence reach "the Racquette River," by a pull of three miles on the Stony Creek. Camp for the night.

THURSDAY. Voyage on the Racquette River of 20 miles to Tupper's Lake. The tourist is here at the last and most charming portion of the region comprised in our present tour ; and here, be he artist or hunter, he will be very willing to pass the remainder of the time which his furlough grants to him. Lough Neah is a continuation of the picturesque waters of Tupper's Lake.

FRIDAY. Tupper's Lake (page 109).

SATURDAY. Tupper's Lake (page 109).

SUNDAY. Tupper's Lake (page 109).

THIRD WEEK.

MONDAY. Tupper's Lake.

TUESDAY. Returning ; retraverse the Racquette River.

WEDNESDAY. Arrive at the Middle Saranac Lake.

THURSDAY. Back to the starting point on the Lower Saranac.

FRIDAY. Regain Lake Champlain at Port Kent, or at Westport.

SATURDAY. Home.

*** If the traveller in this wonderful region be addicted to the rifle, the rod, or the pencil, he may extend his visit with pleasure from three weeks to three months. The Adirondack hills and lakes—another portion of this marvellous wilderness—are not far removed from the Saranac ; and one, two, or more weeks might be spent there with great satisfaction (page 169).

TOUR OF FOUR WEEKS,

To the Great Lakes, via Quebec, Montreal, the St. Lawrence Niagara Falls, &c.

FIRST WEEK.

MONDAY. From New York, via Albany and Troy, to Saratoga.

TUESDAY. Saratoga Springs.
WEDNESDAY. To Montreal, by Railway or Steamer on Lake Champlain.
THURSDAY. Montreal.
FRIDAY. To Quebec.
SATURDAY. Quebec and vicinity, visiting the Falls of Montmorenci, the Chaudiere, &c.
SUNDAY. Quebec.

SECOND WEEK.

MONDAY. Grand Trunk Railway, by Montreal, to Toronto, on Lake Ontario.
TUESDAY. Take the Northern Railway of Canada, 95 miles, to Collingwood, on the Georgian Bay, an arm of Lake Huron.
WEDNESDAY. By Steamer, on Lake Huron (page 43), to the Straits of Mackinac (page 43).
THURSDAY. Mackinac.
FRIDAY. Mackinac.
SATURDAY. Steamer to the Sault St. Marie—the connecting link of the waters of Huron and Lake Superior.
SUNDAY. At the Sault de St. Marie, or the "Soo" as it is familiarly called.

THIRD WEEK.

Voyage on Lake Superior.

FOURTH WEEK.

MONDAY. From the Sault de St. Marie (returning) (Steamer on Lake Huron) to Detroit, Michigan.
TUESDAY. Great Western Railway to Suspension Bridge, Niagara Falls.
WEDNESDAY. Niagara Falls.
THURSDAY. Niagara Falls.
FRIDAY. To Utica Central Railway, or to Binghamton, Erie Route.
SATURDAY. To New York.

TOUR OF FOUR WEEKS,

To the Virginia Springs, Weir's Cave, the Natural Bridge, the Peaks of Otter, &c.

FIRST WEEK.

MONDAY. From New York to Philadelphia (pages 174 and 176).
TUESDAY. Philadelphia to Baltimore (page 179).
WEDNESDAY. Baltimore to Washington City (page 215).
THURSDAY. At Washington City—visit Mount Vernon.
FRIDAY. To Alexandria ; and thence, by the Orange and Alexandria Railway, 88 miles ; and from Gordonsville, on the Virginia Central Road, 64 miles to Staunton.
SATURDAY. Stage or Carriage, 17 miles, to Weir's Cave.
SUNDAY. At Weir's Cave (page 239).

SECOND WEEK.

MONDAY. At Weir's Cave, returning in the afternoon to Staunton.
TUESDAY. Continue journey on the Central Road, to Jackson's River, thence to the Sulphur Springs by Stage. (For Natural Bridge, take Stage at Millboro').
WEDNESDAY. *En route.*
THURSDAY. White Sulphur Springs (page 232).

FRIDAY. White Sulphur Springs.
SATURDAY. White Sulphur Springs.
SUNDAY. White Sulphur Springs.

THIRD WEEK

May be devoted to the other Springs of this Region.

FOURTH WEEK.

Visit the Natural Bridge, 63 miles from the White Sulphur Springs; 12 miles from
Lexington; 36 miles from Lynchburg, on the Virginia and Tennessee Railway,
from Richmond, west (page 236); next, see the Peaks of Otter (page 237), in the
same region. Return home by the Virginia and Tennessee Road, from Lynchburg to Richmond (page 220); thence, by the Great Southern Mail Route to
Washington; or, more agreeably, by the James River and the Chesapeake Bay,
to Baltimore; from Baltimore to Philadelphia; from Philadelphia to New York.

** For landscape beauties not mentioned here in the Spring Region of
Western Virginia, see page 240 and following. This tour might be pleasantly
extended to two or even three months.

TOUR OF FOUR WEEKS,

*From New York, via Boston and Portland, to Quebec and the Saguenay, Montreal, the
Ottawa, and the St. Lawrence, returning by Niagara and Trenton Falls, Sara-
toga Springs, and the Hudson River. Détour of ten days (extra) to the White
Mountains.*

FIRST WEEK.

MONDAY. New York to Boston—see Routes page 61 and following.
TUESDAY. At Boston (page 65).
WEDNESDAY. Boston to Portland, Maine—Routes page 58.
THURSDAY. At Portland (page 58).
FRIDAY. From Portland to Quebec, by the Grand Trunk Railway (pages 35 and 59.)

Détour of Ten Days to White Mountains.

[The White Mountains may be pleasantly visited from this part of our present
Route (in ten extra days), stopping at Gorham, N. H., 91 miles on the way
from Portland, reaching Glen House, 8 miles from Gorham, same day; Crawford House, White Mountain Notch, on Saturday; and so on, as per programme
of SECOND WEEK, in Tours, pages 15 and 16; returning to the Glen House by
the Second Sunday, and resuming journey (from Gorham to Quebec) on Monday following.]

SATURDAY. At Quebec (page 35).
SUNDAY. At Quebec.

SECOND WEEK.

MONDAY. At Quebec, visiting Falls of Montmorenci, of the Chaudiere, of St. Anne,
&c. (page 37).
TUESDAY. Excursion to Saguenay River and back to Quebec, as in Tour (page 18).
WEDNESDAY. Excursion to Saguenay River and back to Quebec.
THURSDAY. Excursion to Saguenay River and back to Quebec.
FRIDAY. From Quebec, by Grand Trunk Railway, or St. Lawrence River, to
Montreal.
SATURDAY. Montreal (page 32).
SUNDAY. Montreal (page 32).

THIRD WEEK.

MONDAY. Excursion up the Ottawa River from Montreal and back (pages 26 to 29).

TUESDAY. Excursion up the Ottawa River from Montreal and back.

WEDNESDAY. Excursion up the Ottawa River from Montreal and back.

THURSDAY. Up the St. Lawrence and Lake Ontario (or by Grand Trunk Railway) to Niagara Falls (page 39).

FRIDAY. Up the St. Lawrence and Lake Ontario (or by Grand Trunk Railway) to Niagara Falls.

SATURDAY. At Niagara Falls (page 161).

SUNDAY. At Niagara Falls.

FOURTH WEEK.

MONDAY. Still at Niagara.

TUESDAY. By Central Railway to Utica.

WEDNESDAY. From Utica, 15 miles, to Trenton Falls.

THURSDAY. At Trenton Falls (page 156), returning to Utica in the evening.

FRIDAY. Journey to and stay at Saratoga Springs (page 149).

SATURDAY. Back to New York, via Troy, Albany, and the Hudson River (page 119).

TOUR OF FOUR WEEKS,

To the Upper Mississippi, via Niagara, Detroit, Chicago, Milwaukee, St. Paul, St. Louis, Louisville, Cincinnati, etc.

FIRST WEEK.

MONDAY. From New York to Niagara by the Erie Railway, 444 miles, or by the Central route, 466 miles—a journey more comfortably made in two days than one, if time serves. By Canandaigua, direct, 439 miles.

TUESDAY. Niagara (page 161).

WEDNESDAY. By the Great Western Railway, 229 miles, to Detroit

THURSDAY. By the Michigan Central road, 284 miles, to Chicago.

FRIDAY. Chicago, Ill. (page 341).

SATURDAY. To Milwaukee by steamer on Lake Michigan, or by railway along shore, 85 miles.

SUNDAY. At Milwaukee, Wis. (page 358).

SECOND WEEK.

Visit to St. Paul, Minnesota, leaving Milwaukee on Monday for Madison, Wis., and thence (circuitously) by railway to Dubuque on the Mississippi, or returning, to Chicago, and thence to Dubuque direct, by the Galena and Chicago route. From Dubuque by steamer up the Mississippi river to St. Paul and the Falls of St. Anthony. Returning by the end of the week (second of the tour) via the river, to St. Louis.

THIRD WEEK.

MONDAY. At St. Louis (page 350).

TUESDAY. By the Ohio and Mississippi Railway, and the New Albany and Salem road to Louisville.

WEDNESDAY. At Louisville, Ky. (page 319).

THURSDAY. At Louisville.

(Another week would permit the traveller to visit the Mammoth Cave very agreeably from this the chief point of détour thither.)

FRIDAY. By railway or steamer on the Ohio river to Cincinnati.

SATURDAY. At Cincinnati, Ohio (page 329).
SUNDAY. At Cincinnati.

FOURTH WEEK.

MONDAY. By railway to Columbus, Ohio (page 333).
TUESDAY. Railway to Zanesville, Ohio (page 334).
WEDNESDAY. To Wheeling, Va.
THURSDAY AND FRIDAY. By the Baltimore and Ohio road to Baltimore, or by the Pennsylvania railway to Philadelphia. Both these noble routes are as magnificent in their pictorial attractions as in their grand extent—each traversing a wide extent of country, replete with every variety of natural beauty. For a description of the numberless notable scenes on the Baltimore and Ohio Railway, see pages 204 and following. For mention of the wonders on the Pennsylvania route, consult page 190.
SATURDAY. To New York.

A WINTER TOUR OF SIX WEEKS,

Visiting the Invalid Resorts of Florida, Savannah and Augusta, Geo., Charleston and Columbia, S. C., Richmond, Va., and Washington City.

FIRST WEEK.

SATURDAY. Leave New York by the steamer of Saturday afternoon, and arrive in Savannah Tuesday morning. Spend the rest of the week in Savannah at the Pulaski House, the Mansion House, or the City Hotel (page 275).

SECOND WEEK.

SATURDAY. Leave Savannah in the steamer for Jacksonville, Pilatka and other places on the St. John's river (pages 265 and 266). Spend the week hereabouts.

THIRD WEEK.

At St. Augustine, on the coast, below the mouth of the St. John's (page 266). St. Augustine, or the "Ancient City," as it is sometimes called, from its venerable age, which exceeds that of any other place in the Union, will tempt the visitor to a long tarry with the social attractions which its fame as an invalid resort has secured. The peculiar natural features of the city and the neighborhood, will also win his particular interest.

FOURTH WEEK.

At St. Augustine.

FIFTH WEEK.

Return to Savannah and take the Georgia Central railway to Augusta (page 277) thence by the South Carolina road to Charleston (page 249).

SIXTH WEEK.

MONDAY. By South Carolina railway from Charleston to Columbia.
TUESDAY. At Columbia (page 257), resuming journey in the afternoon.
WEDNESDAY. *En route.*
THURSDAY. At Richmond, Va. (page 220).
FRIDAY. Arrive at Washington City (page 215).
SATURDAY. To Baltimore in the evening.
SUNDAY. At Baltimore (page 199).
MONDAY. To New York.

BRITISH AMERICA.

———— • ◦ • ————

THE possessions of the British Crown in North America, occupy nearly all the upper half of the Continent; a vast territory, reaching from the Arctic seas, to the domains of the United States; and from the Atlantic to the Pacific Oceans. Of this great region, our present explorations will refer, only to the lower and settled portions, known as the British Provinces—the Canadas, New Brunswick, and Nova Scotia. The rest is for the most part yet a wilderness.

CANADA.

GEOGRAPHY AND AREA. Canada, the largest and most important of the settled portions of the British territory in North America, lies upon all the northern border of the United States, from the Atlantic coast to the waters of Lake Superior and the Mississippi. The two provinces into which it is divided, were formerly known as Upper and Lower Canada, or Canada East and Canada West; and thus, indeed, their differing manners, habits and laws, still virtually divide and distinguish them, though they are now nominally and politically united. The entire length of the Canadian domain from east to west, is between twelve hundred and thirteen hundred miles, with a breadth varying from two to three hundred miles.

DISCOVERY, SETTLEMENT, AND RULERS. The earliest discovery of the Canadas is ascribed to Sebastian Cabot, 1497; the first European settlement was made at St. Croix's Harbour, in 1541, by Jacques Cartier, a French adventurer, who entered and named the river St. Lawrence. In 1608, another and more considerable settlement was made upon the spot now occupied by the city of Quebec. From that period until 1759, the country continued under the rule of France; and then came the capture of Quebec by the English, under General Wolfe, and the transfer within a year thereafter, of all the territory of New France, as the country was at that time called, to the British power, under which it has ever since remained. The mutual disagreement which naturally arose from the conflicting interests and prejudices of the two opposing nationalities, threatened internal trouble from time to time, and finally displayed itself in the overt acts, recorded in history as the rebellion of 1837. It was after these incidents, and as a consequence thereof, that the two sections of the territory were formed into one. This happened in 1840.

GOVERNMENT. Canada is ruled by an executive, holding the title of Governor·

2

General, received from the crown of Great Britain, and by a legislature called the Provincial Parliament. This body consists of an Upper and a Lower House; the members of the one were formerly appointed by the Queen, but now (as fast as those thus placed die) this body is, like the other branch, chosen by the people; each for a term of twelve years.

RELIGION. The dominant religious faith in Lower Canada or Canada East, is that of the Romish Church; while in the upper province the creed of the English Establishment prevails.

LANDSCAPE. The general topography of Upper or Western Canada, is that of a level country, with but few variations excepting the passage of some table heights, extending south-westerly. It is the most fertile division of the territory, and thus to the tourist in search of the picturesque, the least attractive.

The Lower Province, or Canada East, is extremely varied and beautiful in its physical aspect; presenting to the delighted eye a magnificent gallery of charming pictures of forest wilds, vast prairies, hill and rock-bound rivers, rushing waters, bold mountain heights, and all, every where intermingled, and their attractions embellished by intervening stretches of cultivated fields, and rural villages, and villa homes.

MOUNTAINS. The hill ranges of Canada are confined entirely to the lower or eastern province.   The chief lines called the Green Mountains follow a parallel course south-westerly. They lie along the St. Lawrence River, on its southern side, extending from the latitude of Quebec to the Gulf of St. Lawrence.   There is another and corresponding range on the north side of the river, with a varying elevation of about 1,000 feet.   The Mealy Mountains, which extend to Sandwich Bay, rise in snow-capped peaks to the height of 1,500 feet.   The Wotchish Mountains, a short, crescent-shaped group, lie between the Gulf of the St. Lawrence and Hudson's Bay.

RIVERS. Canada has many noble and beautiful rivers, as the St. Lawrence, one of the great waters of the world; the wild, mountain-shored floods of the Ottawa, and the Saguenay; and the lesser waters of the Sorel or Richelieu, the St. Francis, the Chaudière, and other streams.

The St. Lawrence. This grand river which drains the vast inland seas of America, extends from Lake Ontario, 750 miles to the Gulf of St. Lawrence, and thence to the sea. Its entire length including the great chain of lakes by which it is fed, is not less than 2,200 miles. Ships ascend to Quebec, and vessels of 600 tons or more to Montreal.   Its chief affluents are the Saguenay eastward, and the Ottawa on the west.   The width of the St. Lawrence varies from half a mile to four miles; at its mouth it is 100 miles across. It abounds in beautiful islands, of which there is a vast group, near its egress from Lake Ontario, known and admired by all the world, as the "Thousand Isles."

The Thousand Islands. It is a curious speculation to the voyager always, how his steamer is to find its way through the labyrinth of the thousand islands, which stud the broad waters like the countless tents of an encamped army, and ever and anon his interest is aroused up to the highest pitch at the prospective danger of the passage of some angry rapid.   All the journey east, from lake to lake of the great waters, past islands now miles in circuit, and now large enough only for the cottage of Lilliputian lovers, is replete with ever-changing pleasure.

Montreal and Quebec, the chief cities of Canada, are upon the St. Lawrence, while Toronto lies on the shores of Lake Ontario, the continuing waters westward.

The Ottawa River flows 800 miles and enters the St. Lawrence on both sides of the Island of Montreal, traversing in its way Lake Temiscaming, Grand Lake, and others.   Rapids and falls greatly impede the navigation of its waters, but lend to them wonderful beauty. It is a wild forest region; that of the Ottawa, but little occupied heretofore by others than the rude lumbermen; though numerous settlements are now springing up, and its agricultural capacities are being developed.

The Committee on Railways of the House of Assembly of the Province, in its report, thus speaks of this river:—

"At the head of the lake the Blanche River falls in, coming about 90 miles from the north. Thirty-four miles farther down the lake it receives the Montreal River, coming 120 from the north-west. Six miles lower down, on the east or Lower Canada bank it receives the Keepawa-sippi, a large river which has its origin in a lake of great size, hitherto but partially explored, and known as Lake Keepawa. This lake is connected with another chain of irregularly shaped lakes, from one of which proceeds the river Du Moine, which enters the Ottawa about 100 miles below the mouth of the Keepawa-sippi; the double discharge from the same chain of lakes in opposite directions presents a phenomenon similar to the connection between the Orinoco and Rio Negro in South America. The Keepawa-sippi has never been surveyed, but on a partial survey of the lake from which it proceeds, it was found flowing out with a slow and noiseless current, very deep, and about 300 feet in width; its middle course is unknown, but some rafts of timber have been taken out a few miles above the mouth. It is stated in the report from which we quote, that there is a cascade at its mouth 120 feet in height; this is a fable; the total descent from the lake to the Ottawa may be 120 feet, but there is no fall at the mouth of the river.

"From the Long Sault at the foot of Lake Temiscaming, 233 miles above Bytown, and 360 miles from the mouth of the Ottawa, down to Deux Joachim Rapids, at the head of the Deep River, that is for 89 miles, the Ottawa, with the exception of 17 miles below the Long Sault, and some other intervals, is not at present navigable except for canoes. Besides other tributaries in the interval, at 197 miles from Bytown, now called Ottawa, it receives on the west side the Mattawan, which is the highway for canoes going to Lake Huron, by Lake Nipissing. From the Mattawan the Ottawa flows east by south to the head of Deep River reach, nine miles above which it receives the river Du Moine from the north.

"From the head of Deep River, as this part of the Ottawa is called, to the foot of Upper Allumettee Lake, two miles below the village of Pembroke, is an uninterrupted reach of navigable water, 43 miles in length. The general direction of the river in this part is south-east. The mountains along the north side of Deep River are upwards of 1,000 feet in height, and the many wooded islands of Allumettes lake render the scenery of this part of the Ottawa magnificent and exceedingly picturesque—far surpassing the celebrated lake of the Thousand Islands on the St. Lawrence.

"Passing the short rapid of Allumettes, and turning northward, round the lower end of Allumettes Island, which is 14 miles long, and 8 at its greatest width, and turning down south-east through Coulonge Lake, and passing behind the nearly similar islands of Calumet, to the head of the Calumet Falls, the Ottawa presents, with the exception of one slight rapid, a reach of 50 miles of navigable water. The mountains on the north side of Coulonge lake, which rise apparently to the height of 1,500 feet, add a degree of grandeur to the scenery, which is in other respects beautiful and varied. In the Upper Allumettes Lake, 115 miles from Ottawa, the river receives from the west the Petawawee, one of its largest tributaries. This river is 140 miles in length, and drains an area of 2,200 square miles. At Pembroke, 9 miles lower down on the same side, an inferior stream, the Indian River, also empties itself into the Ottawa.

"At the head of Lake Coulonge, the Ottawa receives from the north, the

Black River, 130 miles in length, draining an area of 1,120 miles; and 9 miles lower, on the same side, the river Coulonge, which is probably 160 miles in length, with a valley of 1,800 square miles.

"From the head of the Calumet Falls to Portage du Fort, the head of the steamboat navigation, a distance of 8 miles, are impassable rapids. Fifty miles above the city, the Ottawa receives on the west the Bonnechère, 110 miles in length, draining an area of 980 miles. Eleven miles lower, it receives the Madawaska, one of its greatest feeders, a river 210 miles in length, and draining 4,100 square miles.

"Thirty-seven miles above Ottawa there is an interruption to the navigation, caused by three miles of rapids and falls, to pass which a railroad has been made. At the foot of the rapids, the Ottawa divides among islands into numerous channels, presenting a most imposing array of separate falls.

"Six miles above Ottawa begin the rapids terminating in the Ottawa Chaudière Falls, which, inferior in impressive grandeur to the Falls of Niagara, are perhaps more permanently interesting, as presenting greater variety.

"The greatest height of the Chaudière Falls is about 40 feet. Arrayed in every imaginable variety of form in vast dark masses, in graceful cascades, or in tumbling spray, they have been well described as a hundred rivers struggling for a passage. Not the least interesting feature which they present is the Lost Chaudière, where a body of water greater in volume than the Thames at London, is quietly sucked down, and disappears under ground.

"At the city of Ottawa the river receives the Rideau from the west, running a course of 116 miles, and draining an area of 1,350 square miles."

The city of Ottawa, on the banks of the river, is thought to be excelled in the beauty of its position, only by Quebec, on the St. Lawrence. From Barrack Hill here, the wide panorama includes the Falls of the Chaudière, the Suspension Bridge, which connects the upper and lower provinces, the islanded stretch of the river above, and of the far-away mountain ranges.

The Rideau Falls, near the mouth of the Rideau, just below the city of Ottawa, is a charming scene.

"A mile lower it receives, from the north, its greatest tributary, the Gatineau, which, with a course probably of four hundred and twenty miles, drains an area of twelve thousand square miles. For about two hundred miles the upper course of this river is in the unknown northern country. At the farthest point surveyed, two hundred and seventeen miles from its mouth, the Gatineau is still a noble stream, a thousand feet wide, diminished in depth but not in width.

"Eighteen miles lower down, the Rivière au Lièvre enters from the north, after running a course of two hundred and sixty miles in length, and draining an area of four thousand one hundred miles. Fifteen miles below it, the Ottawa receives the North and South Nation Rivers on either side, the former ninety-five, and the latter a hundred miles in length. Twenty-two miles further, the River Rouge, ninety miles long, enters from the north. Twenty-one miles lower, the Rivière du Nord, a hundred and sixty miles in length, comes in on the same side; and lastly, just above its mouth, it receives the River Assumption, which has a course of a hundred and thirty miles.

"From Ottawa the river is navigable to Grenville, a distance of fifty-eight miles, where the rapids that occur for twelve miles are avoided by a succession of canals. Twenty-three miles lower, at one of the mouths of the Ottawa, a single lock, to avoid a slight rapid, gives a passage into Lake St. Louis, an expansion of the St. Lawrence, above Montreal.

"The remaining half of the Ottawa's waters find their way to the St. Lawrence, by passing in two channels, behind the Island of Montreal and the Isle Jesus, in a course of thirty-one miles. They are interrupted with rapids, still it is by one of them that all the Ottawa lumber passes to market. At Bout de l'Isle, therefore, the Ottawa is finally merged in the St. Lawrence, a hundred and thirty miles below from the city of Ottawa.

"The most prominent characteristic of the Ottawa is its great volume. Even above the town, where it has to receive tributaries equal to the Hudson, the Shannon, the Thames, the Tweed, the Spey, and the Clyde, it displays, when unconfined, a width of half-a-mile of strong boiling rapid, and when at the highest, while the north waters are passing, the volume, by calculated approximation, is fully equal to that passing Niagara, that is, double the common volume of the Ganges.

"Taking a bird's-eye view of the valley of the Ottawa, we, see spread out before us a country equal to eight times the State of Vermont, or ten times that of Massachusetts, with its great artery, the Ottawa, curving through it, resembling the Rhine in length of course, and the Danube in magnitude.

"This immense region overlies a variety of geological formations, and presents all their characteristic features, from the level uniform surface of the silurian system, which prevails along a great extent of the Ottawa, to the rugged and romantic ridges in the metamorphic and primitive formations which stretch far away to the north and north-west.

"As far as our knowledge of the country extends, we find the greater part of it covered with a luxuriant growth of red and white pine timber, making the most valuable forests in the world, abundantly intersected with large rivers, fitted to convey the timber to market, when manufactured.

"The remaining portion of it, if not so valuably wooded, presents a very extensive and advantageous field for settlement. Apart from the numerous townships already surveyed and partly settled, and the large tracts of good land interspersed throughout the timber country, the great region on the upper course of the western tributaries of the Ottawa, behind the red pine

country, exceeds the State of New Hampshire in extent, with an equal climate and superior soil. It is generally a beautiful, undulating country, wooded with a rich growth of maple, beech, birch, elm, &c., and watered with lakes and streams, affording numerous mill-sites, and abounding in fish. Flanking on the one side the lumbering country, which presents an excellent market for produce, and adjoining Lake Huron on the other, the situation, though comparatively inland, is highly advantageous. In the diversity of resources the Ottawa country presents unusual attractions alike to agricultural and commercial enterprise."

Routes from Montreal up the Ottawa. —Steamers run daily, during the summer months, between Montreal and Ottawa, and Kingston and Ottawa, via the Rideau Canal. Above Ottawa the traveller may proceed by carriage or by stage, nine miles, to the village of Aylmer, and thence by steamer to the Chats; thence by railway, two miles; then again by steamer to the Portage du Fort: now, wagons for awhile, and then again a steamer to Pembroke, and yet another from thence to Deux Joachims; afterwards he must canoe it. The Ottawa may also be reached by Railway direct, from Prescott on the St. Lawrence, to Ottawa City.

The Saguenay. The journey up this beautiful river may be made semi-weekly, by steamer from Quebec. The course of the Saguenay—between lofty and precipitous heights; and in its upper part, amid rushing cataracts, is 126 miles from Lake St. John to the St. Lawrence, which it enters 140 miles below Quebec. Large ships ascend 40 miles.

In the trip from Quebec to the Saguenay beauties there, are many interesting points to be noted in the preceding journey of 120 miles down the St. Lawrence; the ancient-looking settlements on its banks, and the not less picturesque *habitans* of the country. A day's sail lands the voyager at River du Loup, where he passes the night on board his steamer, waiting for

The Mouth of the Saguenay.

the following morning to resume his journey.

The Saguenay is a perfectly straight river, with grand precipices on either side. It has neither windings nor projecting bluffs, or sloping banks or sandy shores like other rivers, nor is its stern, strange aspect varied by either village or villa. "It is," says a voyager thither, "as if the mountain range had been cleft asunder, leaving a horrid gulf of 60 miles in length and 4,000 feet in depth, through the grey mica schist, and still looking fresh and new. One thousand five hundred feet of this is perpendicular cliff, often too steep and solid for the hemlock or dwarf-oak to find root; in which case, being covered with colored lichens and moss, their fresh-looking fractures often appear, in shape and color, like painted fans, and are called the pictured rocks. But those parts more slanting are thickly covered with stunted trees, spruce, and maple, and birch growing wherever they can find crevices to extract nourishment; and the bare roots of the oak, grasping the rock, have a resemblance to gigantic claws. The bases of these cliffs lie far under water, to an unknown depth. For many miles from its mouth no soundings have been obtained with two thousand feet of line; and, for the entire distance of 60 miles, until you reach Ha Ha Bay, the largest ships can sail, without obstruction from banks or shoals, and, on reaching the extremity of the bay, can drop their anchor in 30 fathoms. The view up this river is singular in many respects; hour after hour, as you sail along, precipice after precipice unfolds itself to view, as in a moving panorama; and you sometimes forget the size and height of the objects you are contemplating, until reminded by seeing a ship of one thousand tons lying like a small pinnace under the towering cliff to which she is moored; for, even in these remote and desolate regions, industry is at work, and, although you cannot much discern it, saw-mills have been built on some of the tributary streams which fall into the Saguenay. But what strikes one most, is the absence of beach or strand, for except in a few places where mountain torrents, rushing through gloomy ravines, have washed down the *detritus* of the hills, and formed some alluvial land at the mouth, no coves, nor creeks, nor projecting rocks are seen in which a boat could find shelter, or any footing be obtained. The characteristic is a steep wall of rock rising abruptly from the water; a dark and desolate region, where all is cold and gloomy; the mountains hidden with driving mist, the water black as ink, and cold as ice. No ducks nor sea-gulls sitting on the water, or screaming for their prey. No hawks nor eagles soaring overhead, although there is abundance of what might be called 'Eagle Cliffs.' No deer coming down to drink at the streams, no squirrels nor birds to be seen among the trees. No fly on the water, nor swallow skimming over the surface. It reminds you of

'That lake whose gloomy shore
Sky-lark never warbled o'er.'

Two living things you may see, but these are cold-blooded animals; you may see the cold seal, spreading himself upon his clammy rock, watching for his prey. You may see him make his sullen plunge into the water, like to the Styx for blackness. You may see him emerge again, shaking his smooth oily sides, and holding a huge living salmon writhing in his teeth; and you may envy the fellow faring so sumptuously, until you recollect that you have just had a hearty breakfast of fresh grilled salmon yourself, and that you enjoyed it as much as your fellow creature is now enjoying his raw morsel. And this is all you see for the first twenty miles, save the ancient settlement of Tadousac at the entrance, and the pretty cove of L'Ance a l'Eau, which is a fishing station.

" Now you reach Cape Eternite, Cape Trinité, and many other overhanging cliffs, remarkable for having such clean fractures, seldom equalled for boldness and effect, which create constant apprehensions of danger, even in a calm, but if you happen to be caught in a thunder-storm, the roar, and darkness, and flashes of lightning are perfectly frightful. At last you terminate your voyage at Ha Ha Bay; that is, smiling or laughing bay, in the Indian tongue, for you are perfectly charmed and relieved to arrive at a beautiful spot, where you have sloping banks, a pebbly shore, boats, and wherries, and vessels riding at anchor; birds and animals, a village, a church, French Canadians, and Scottish Highlanders."

After duly enjoying the pleasant " let down " from the high tragic tone of the landscape you have been so long gazing upon and wondering at, formed in the comparatively pastoral character of this upper region of the Ottawa, you return to your steamer, and descending the stern and solemn river, come again, at night-fall, to the River du Loup, from whence you started in the morning. This is the second day of your journey, and on the third you are back once more in Quebec.

SPRINGS.

The Caledonia Springs, a place of much resort, are at the village of Caledonia, 72 miles from Montreal. Leave Montreal by the Lachine railway, and take the steamer to Carillon. At Point Fortune, opposite Carillon, on the other side of the Ottawa, take stage to the springs, arriving the same evening. Fare generally $2.25. Board at the springs from $1.25 to 75 cents per day.

Plantagenet Springs. From Montreal to Point Fortune, as in the route to the Caledonia springs; and thence by stage, arriving same evening. Distance 88 miles. Fare about $1. The consumption of the "Plantagenet water" is said to be very great.

The St. Leon Springs are at the village of St. Leon, on the Riviere du Loup, between Montreal and Quebec; 26 miles by stage from Three Rivers, a landing of the St. Lawrence steamers.

St. Catharine's, Canada West, on the Great Western Railway, 11 miles from Niagara Falls, and 32 miles from Hamilton. See St. Catharine's in route from Montreal to Niagara via the St. Lawrence.

WATERFALLS IN CANADA.

Niagara. See chapter on the state of New York.

Falls of Montmorenci. See Quebec.

The Chaudiere Falls on the Ottawa. See Ottawa river.

The Chaudiere Falls, Quebec. See City of Quebec.

The Rideau Falls. See Ottawa river.

The Falls of Shawanegan are on the river St. Maurice, 25 miles from Three Rivers, on the St. Lawrence river, between Montreal and Quebec. The St. Maurice, 186 feet in breadth at this point, makes a perpendicular descent of about 200 feet. The imposing character of this scene is, as yet, but little known. Between the Falls and the town of Three Rivers, the St. Maurice affords excellent fishing.

St. Anne's Falls are 24 miles below Quebec. See Quebec.

RAILWAYS.

The Grand Trunk connects Montreal with Quebec, and each with Portland in Maine. From Montreal it follows the upper shore of the St. Lawrence and of Lake Ontario to Toronto, and thence continues westward, across the peninsula of Canada West, to the State of Michigan, at Port Sarnia, on the southern extremity of Lake Huron. The western extremity of the route is not yet completed. It connects with routes to Niagara Falls, with the line of the Great Western Railway, and with the routes Mississippiwards.

The Great Western Railway extends from Niagara Falls, 229 miles west to Detroit, Michigan, connecting with the Michigan central route for Chicago, &c.

The Montreal and New York road extends from Montreal 67 miles to Plattsburg, and is a part of the route from Montreal to New York.

The Champlain and St. Lawrence extends from Montreal, 44 miles, to Rouse's Point on Lake Champlain, thence to New York, Boston, &c.

The Northern Railway of Canada extends 94 miles from Toronto on Lake Ontario to Collingwood on the Georgian Bay, Lake Huron. A part of a pleasant route from New York to Lake Superior.

The Ottawa and Prescott Railway extends from Prescott (opposite Ogdensburg) on the St. Lawrence, 54 miles to Ottawa, on the Ottawa river.

The Hamilton and Toronto road extends 38 miles from Toronto to Hamilton, connecting the Grand Trunk and the Great Western routes.

The Coburg and Peterboro' Railway, 28 miles from Peterboro' to Coburg, on the line of the Grand Trunk, between Montreal and Toronto.

Many other routes are either in progress, or in contemplation — Canada vying with the "States" in this field of enterprise.

MONTREAL.

Montreal may be reached daily from New York in from 15 to 18 hours, by the Hudson river or Harlem railway to Albany; thence by railway to Whitehall; thence by steamer on Lake Champlain, or railway through Vermont to Plattsburg; thence by the Montreal and New York road. From Boston via Albany, or other routes to Lake Champlain, &c.; or, via Portland and the Grand Trunk railway.

Hotels. The Donegana, Notre-Dame street; the St. Lawrence, Great St. James street, a fine house, centrally located; the Ottawa, Great St. James street; and the Montreal House, Custom House square, and opposite the Custom House. Besides these leading establishments, there are many other comfortable houses and cafés, where travellers of all ranks and classes may be lodged and regaled according to the varied humors of their palates and their purses. We forbear to name too many, lest the bewildered stranger should, in the very abundance of the good things placed before him, starve while choosing.

Montreal, the most populous city in British North America, is picturesquely situated at the foot of the Royal Mountain, from which it takes its name, upon a large island at the confluence of

Montreal.

the Ottawa and St. Lawrence, which, both in fertility and cultivation, is considered the garden of Canada East. The main branch of the Ottawa, which is the timber highway to Quebec, passes north of Montreal Island, and enters the St. Lawrence about 18 miles below the city; about one-third of its waters are, however, discharged into Lake St. Louis, and joining but not mingling at Caughnawaga, the two distinct bodies pass over the Sault St. Louis and the Lachine Rapids—the dark waters of the Ottawa washing the quays of Montreal, while the blue St. Lawrence occupies the other shore. Nor do they merge their distinctive character until they are several miles below Montreal. The quays of Montreal are unsurpassed by those of any city in America; built of solid limestone, and uniting with the locks and cut stone wharves of the Lachine Canal, they present for several miles a display of continuous masonry, which has few parallels. Unlike the levees of the Ohio and the Mississippi, no unsightly warehouses disfigure the

river side. A broad terrace, faced with gray limestone, the parapets of which are surmounted with a substantial iron railing, divides the city from the river throughout its whole extent.

The people in Montreal number nearly 75,000, and the population steadily increasing. The houses in the suburbs are handsomely built in the modern style, and mostly inhabited by the principal merchants. Including its suburbs, of which it has several, the city stretches along the river for two miles from s. w. to n. e., and, for some distance, extends between one and two miles inland. It was formerly surrounded by a battlemented wall; but this having fallen into decay, it is now entirely open. St. Paul st. the chief commercial thoroughfare, extends along the river the whole length of the city. Notre-Dame-street is the fashionable promenade.

The French Cathedral. Of the public buildings, the most remarkable is the Roman Catholic Cathedral, Place d'Armes, constructed in the Gothic

2*

style, with a length of 255½ feet, and a breadth of 134½ feet. It has two towers, each of which has a noble elevation of two hundred and twenty-five feet. The view from these towers—embracing the city and its suburbs, the river, and the surrounding country — is exceedingly beautiful. The principal window of the Cathedral is 64 feet high and 32 broad. Of the vastness of the interior of this edifice an idea may be formed from the fact that it is capable of accommodating eight or nine thousand persons. This immense assembly may, by numerous outlets, disperse in five or six minutes.

The Seminary of St. Sulpice, adjoining the Cathedral, is 132 feet long, and 29 deep, and is surrounded by spacious gardens and court yard.

The Bank of Montreal and the City Bank, the first a fine example of Corinthian architecture, are on the opposite sides of the square, or Place d'Armes.

St. Patrick's Church (Catholic) oc-

cupies a commanding position at the west end of Lagauchetiere street.

The Bishop's Church (Catholic) is a very elegant structure in St. Dennis street.

The remaining Catholic churches are the Recollet, in Notre Dame street, the Bonsecours, near the large market, and the St. Mary's, in Griffintown. There are also chapels attached to all the Nunneries, in some of which excellent pictures may be seen.

Nunneries. The Grey Nuns, in Foundling street, was founded, in 1692, for the care of lunatics and children. The Hotel Dieu was established in 1644, for the sick generally. The Black, or congregational nunnery, in Notre Dame street, dates from 1659. The Sisterhood, at this third and last of the conventual establishments of Montreal, devote themselves to the education of young persons of their own sex.

The stranger desirous of visiting either of the nunneries should apply to the Lady Superior for admission, which is seldom refused.

The Protestant churches worthy of notice are St. Andrew's Church, in Bever Hall, a beautiful specimen of Gothic architecture, being a close imitation of Salisbury Cathedral, in England, though of course on a greatly reduced scale. This, with St. Paul's Church, in St. Helen street, are in connection with the Established Church of Scotland. The Episcopalian churches are, the beautiful new edifice—Christ Church Cathedral, St. George's Church, in St. Joseph street, St. Stephen's, in Griffintown, Trinity, in St. Paul street, and St. Thomas', in St. Mary street. Various other denominations of Christians have churches—the Wesleyans, a large and very handsome building, in St. James' street, and also others in Griffintown and Montcalm street; the Independents, two, one in Gosford street, and one in Radegonde street. This last was the scene of the sad riot and loss of life on the occasion of Gavazzi's lecture in 1852. The Free Church has also two places of worship, one in Coté street, and one in St. Gabriel street; besides these, there are the American and the United Presbyterian, the Baptist, and the Unitarian Churches; a small Jewish Synagogue, the last named being classical in design.

Directly opposite the city is the wharf of the New York and St. Lawrence Railway Company. Below Nun's Island are seen the gigantic piers of the Tubular Bridge, a wonderful structure, which is to span the great St. Lawrence.

The Bonsecours Market is an imposing Doric edifice, erected at a cost of $280,000. In one of the upper stories are the Offices of the Corporation and Council Chamber, and a concert or ball-room capable of seating 4,000 people. The view from the dome of this structure, overlooking the river and St. Helen's Isle, are well worth the seeing.

At the head of Place Jacques Cartier there is a column erected to the memory of the naval hero, Lord Nelson.

The Court House is one of the most striking of the architectural specialities of the city.

The Post Office is in Great James' street.

The Custom House is a neat building on the site of an old market-place, between St. Paul street and the river.

The Merchants' Exchange and Reading-Room are in St. Sacrament street. The latter is a large and comfortable room, well supplied with newspapers and periodicals, English and American, all at the service of the stranger, when properly introduced.

The General Hospital and St. Patrick's are in Dorchester street; the latter, however, at the west end of the town.

M'Gill's College is beautifully situated at the base of the mountain. The High School department of the college is in Belmont street.

The city also possesses an English University, chartered in 1821; and many other institutions for the promotion of learning—French and English seminaries, a royal grammar-school, with parochial, union, national, Sunday, and other public schools. It has numerous societies for the advancement of religion, science, and industry; and several public libraries.

The Water-Works, a mile or so from the city, are extremely interesting for their own sake, and for the fine view of the neighborhood to be seen thence.

The Mount Royal Cemetery is two miles from the city, on the northern slope of the mountain. From the high road round its base, a broad avenue through the shaded hill-side gradually ascends to this pleasant spot.

There are other romantic burying-grounds, both of the Catholic and the Protestant population, in the vicinity of Montreal, and other scenes which the visitor should enjoy—pleasant rides all about, around the mountain and by the river, before he bids good-bye to the Queen City of Canada.

QUEBEC.

Quebec may be pleasantly reached from New York, via Boston to Portland, Maine, and thence 317 miles by the Grand Trunk Railway, total distance, by this route, from New York to Quebec, 650 miles; or, from New York by the Hudson River Railway or steamboats; or by the Harlem Railway to Albany, thence to Whitehall, thence on Lake Champlain to Plattsburg, thence by the Montreal and New York Railway to Montreal, and from Montreal by steamer down the St. Lawrence, or by the Grand Trunk Railway. Distance by railway, from Montreal to Quebec, 168 miles. There are other railway routes from Boston to Quebec, via Albany, or via the Vermont lines to Lake Champlain and Montreal.

Hotels. The principal hotels are the Russell and the Clarendon. The Russell is in Palace street, Upper Town, and is the favorite headquarters of American tourists, while the English establish themselves, generally, at the Clarendon on St. Louis street. After these come Swords', corner of Haldimand and St. Louis streets, U. T.; City Hotel, 1 Garden street, U. T., and many others.

Quebec is the capital of United Canada, and, after Montreal, the most populous city in British North America. It is upon the left bank of the St. Lawrence river, and some 340 miles from the Ocean.

The city was founded in 1608, by the geographer, Champlain. It fell into the possession of the British in 1619, but was restored three years later. The English made an unsuccessful attempt to regain possession in 1690, but it did not finally come into their hands until taken by General Wolfe, in 1759.

The city is divided into two sections, called the Upper and the Lower Towns; the Upper Town occupying the highest part of the promontory, which is surrounded by strong walls and other fortifications; and the Lower Town, being built around the base of Cape Diamond. The latter is the business quarter.

The Citadel, a massive defence, crowning the summit of Cape Diamond, covers about 40 acres with its numerous buildings. Its impregnable position makes it perhaps the strongest fortress on this continent; and the name of the "Gibraltar of America" has been often given to it not inaptly. The walls of the Citadel are entered by five gates. That, called the Palace Gate, leads to the Ashley Barracks, which have accommodation for 2,000 troops. The St. Louis Gate, on the south-west, gives access to the Plains of Abraham, the scene of Wolfe's victory and death, in 1759, and of the death of Montgomery, in December, 1775. The Prescott Gate is the only entrance on the St. Lawrence side of the fortress.

The view from the Citadel is remarkably fine, taking in, as it does, the opposite banks of the great river for nearly half a hundred miles up and down. The promenade here, on the ramparts above the esplanade, is charming. In the public garden, on Des Carrierres street, there is an obelisk to the memory of Wolfe and Montcalm.

The Parliament House. Among the chief public edifices of Quebec is the New Parliament House, which supplies the place of the building destroyed by fire in 1854.

The Roman Catholic Cathedral was erected under the auspices of the first Bishop of Quebec, and was consecrated in 1666. It is 216 feet long, and 180 feet in breadth.

The Ursuline Convent and the Church of St. Ursula are agreeable buildings, encompassed by pleasant gardens. This establishment was founded in 1639, and holds a high position in the public esteem. It contains a Superior, fifty nuns, and six novices, who give instruction in reading, writing, and needle-work. The convent was

The Citadel, Quebec.

destroyed by fire in 1650, and again in 1686. The remains of the Marquis de Montcalm are buried here.

The Artillery Barracks form a range of stone buildings 500 feet in length.

Durham Terrace is the site of the old castle of St. Louis, which was entirely consumed by fire in 1834.

The English Protestant Cathedral, consecrated in 1804, is one of the finest modern edifices of the city.

St. Andrew's Church, in St. Anne street, is in connection with the Scotch Establishment. The Methodists have a chapel in St. Stanislaus's street, and another in St. Louis suburb, called the Centenary Chapel.

The Lower Town. The passage from the Upper to the Lower Town is by the Prescott gate, though foot passengers take the shorter way known as the Break-neck Stairs. There will be found the Exchange, the Banks, and other commercial establishments.

The Plains of Abraham may be reached via the St. Louis Gate, and the counterscarp on the left, leading to the glacis of the citadel; hence towards the right; approaching one of the Martello Towers, where a fine view of the St. Lawrence opens. A little beyond, up the right bank, is the spot where General Wolfe fell on the famous historic ground of the Plains of Abraham. It is the highest ground, and is surrounded by wooden fences. Within an enclosure lower down is a stone well, from which water was brought to the dying hero.

Wolfe's Cave, the spot where *Montgomery* was killed, and other scenes, telling tales of the memorable past, will be pointed out to the traveller in this neighborhood.

The Mount Hermon Cemetery is about three miles from the city, on the south side of the St. Louis Road. The grounds are 32 acres in extent, sloping irregularly but beautifully down the precipices which overhang the St. Lawrence. They were laid out by the late Major Douglass, of the U. S. Engineers, who had previously displayed his skill and taste in the arrange-

ments of the Greenwood Cemetery, near New York.

Lorette. To see Lorette may be made the motive of an agreeable excursion from Quebec, following the banks of the St. Charles.

Lake St. Charles is four miles long, and one broad. It is divided by projecting ledges into two parts. It is a delightful spot, in its natural attractions, and in the fine sport it affords to the angler.

The Falls of Montmorenci, eight miles distant, are among the chief delights of the vicinage of Quebec. The river here is 60 feet wide, and the descent of the torrent 250 feet.

"The effect of the view of these falls upon the beholder is most delightful. The river, at some distance, seems suspended in the air, in a sheet of billowy foam, and contrasted, as it is, with the black frowning abyss into which it falls, it is an object of the highest interest. The sheet of foam, which first breaks over the ridge, is more and more divided as it plunges and is dashed against the successive layers of rock, which it almost completely veils from view; the spray becomes very delicate and abundant, from top to bottom, hanging over, and revolving around the torrent, till it becomes lighter and more evanescent than the whitest fleecy clouds of summer, than the finest attenuated web, than the lightest gossamer, constituting the most airy and sumptuous drapery that can be imagined. Yet, like the drapery of some of the Grecian statues, which, while it veil, exhibits more forcibly the form beneath, this does not hide but exalts the effect produced by this noble cataract.

"Those who visit the falls in the winter, see one fine feature added to the scene, although they may lose some others. The spray freezes, and forms a regular cone, of 100 feet and upwards in height, standing immediately at the bottom of the cataract, like some huge giant of fabulous notoriety."

The extraordinary formation called

Falls of Montmorenci.

the Natural Steps, will not fail to interest the visitor at Montmorenci.

The Falls of St. Anne, in the river
St. Anne, 24 miles below Quebec, are
in a neighborhood of great picturesque
beauty. Starting from the city in the
morning betimes, one may visit Montmorenci nicely, and proceed thence
the same evening to St. Anne. Next
morning after a leisurely survey of these
cascades, there will be most of the day
left to get back, with any *détours* that
may seem desirable, to Quebec.

The Falls of the Chaudiere are
reached via Point Levi. The rapid
river plunges over a precipice of 130
feet, presenting very much the look of
boiling water, from whence its name of
chaudiere or caldron. The cataract is
broken into three separate parts by the
intervention of huge projecting rocks,
but it is reunited before it reaches the
basin beneath.

We take our leave of this venerable
city, its unique natural beauties, and its
winning stories, with the remembrance
of some of the impressions it made upon
Professor Silliman, when he visited it
years ago:—"Quebec," he writes, " at
least for an American city, is certainly
a very peculiar place. A military town
—containing about 20,000 inhabitants
—most compactly and permanently
built—environed, as to its most important parts, by walls and gates—and defended by numerous heavy cannon—

garrisoned by troops having the arms, the costume, the music, the discipline of Europe—foreign in language, features, and origin, from most of those whom they are sent to defend—founded upon a rock, and in its highest parts overlooking a great extent of country—between 300 and 400 miles from the ocean—in the midst of a great continent, and yet displaying fleets of foreign merchantmen in its fine capacious bay—and showing all the bustle of a crowded seaport—its streets narrow, populous, and winding up and down almost mountainous declivities—situated in the latitude of the finest parts of Europe—exhibiting in its environs the beauty of an European capital—and yet in winter smarting with the cold of Siberia—governed by a people of different language and habits from the mass of the population—opposed in religion, and yet leaving that population without taxes, and in the full enjoyment of every privilege, civil and religious."

Toronto, on Lake Ontario, and the line of the Grand Trunk Railway; from Quebec, 501 miles; from Montreal, 333 miles; from Hamilton, 38 miles; from Niagara Falls, 81 miles. For description, see *Toronto*, in route from Montreal to Niagara, via the St. Lawrence.

Kingston, at the foot of Lake Ontario, on the St. Lawrence, and on the line of the Grand Trunk Railway; from Quebec, 341 miles; from Montreal, 173 miles; from Toronto, 160 miles. See *Kingston* in route from Montreal to Niagara.

Hamilton, near the eastern terminus of the Great Western Railway, at the head of Lake Ontario; from Quebec, 539 miles; from Montreal, 371 miles; from Toronto, 38; from Niagara, 43. See route from Montreal to Niagara.

London is a prosperous town, midway on the line of the Great Western Railway, in its traverse of the peninsula of Canada West from Lake Ontario to Lake Huron. Distant from Niagara Falls, 119 miles, west; from Hamilton, 76 miles; from Toronto, 114 miles; from Montreal, 447 miles; from Quebec,

615 miles. In 1820, the present site of London was a wilderness, occupied by the savages and the wild deer; now its population exceeds 12,000. Like Hamilton, Toronto, and all the growing towns of Canada, it is well built, upon wide streets, and with elegant and substantial architecture.

MONTREAL TO NIAGARA FALLS; UP THE ST. LAWRENCE RIVER AND LAKE ONTARIO.

The traveller may go from Montreal to Niagara, either by steamer on the St. Lawrence, or by the Grand Trunk Railway, 333 miles to Toronto on Lake Ontario. At Toronto he may cross the western end of the .ake to the town of Niagara, and thence reach the Falls by the Erie and Ontario Railway 14 miles long; or he may go less directly, by water or by rail to Hamilton, and thence by rail again to the Falls.

Up THE ST. LAWRENCE FROM MONTREAL. See rivers of Canada for general mention of the St. Lawrence.

Lachine. From Montreal the traveller will proceed nine miles to Lachine by railway, avoiding the rapids which the steamers sometimes descend. At Lachine is the residence of Sir George Simpson, Governor of the Hudson's Bay Company, and of the officers of this, the chief post of that corporation. It is from this point that the orders from head-quarters in London are sent to all the many posts throughout the vast territory of the Company; and near the end of April each year a body of trained *voyageurs* set out hence in large canoes, called *maîtres canots*, with packages and goods for the various posts in the wilderness. Two centuries ago, the companions of the explorer Cartier on arriving here, thought they had discovered a route to China, and expressed their joy in the exclamation of La Chine! Hence, the present name, or so at least says tradition. A costly canal overcomes the obstruction of the rapids at Lachine.

The Village of the Rapids; or, Caughnawaga. An Iroquois settlement lies opposite Lachine, at the outlet of the expansion of the river called

Lake St. Louis. The Indians at Caughnawaga, subsist chiefly by navigating barges and rafts down to Montreal, and in winter, by a trade in moccasins; snow-shoes, &c. They are mostly Roman Catholics, and possess an elegant church.

Lake St. Louis. The brown floods of the Ottawa assist in forming this great expanse of the St. Lawrence. They roll unmixed through the clearer water of the great river. On the northern shore of Lake St. Louis is the island of Montreal, 30 miles long. At the western extremity is *Isle Perrot.* The *Cascade Rapids* separate the expanse just passed from Lake St. Francis. The Beauharnois Canal here is 11¼ miles in length, and has nine locks.

Lake St. Francis, into which the voyager now enters, extends 40 miles. Midway on the right, is the village of LANCASTER, where a pile of stones or *cairn* has been thrown up in honor of Sir John Colborne, formerly Governor-General of Canada, now Lord Seaton. Leaving Lake St. Francis, we pass the passage of the celebrated *Long Sault* rapids. Here, too, is the *Cornwall Canal,* 11½ miles in length, with 7 locks of noble size.

Cornwall is a pleasant town, formerly called "Pointe Maline," in memento of the labor of ascending the river at this point.

The Village of St. Regis lies across from Cornwall. It forms the boundary between Canada and the State of New York, and also intersects the tract of land occupied by the 1,000 Iroquois, American and British, who dwell here.

Dickenson Landing is at the head of the Cornwall canal, within the space of the 38 miles which follow to Prescott; the villages of MOULINETTE, MARIA TOWN, and MATILDA, are successively passed.

The Battle Field of Chryseler's Farm, where the Americans met a defeat in the last war, lies a little above Maria Town.

Prescott is rapidly recovering its prestige lost when the construction of the Rideau Canal won its trade away to Kingston; for now a railway from New York approaches it at Ogdensburg, and another connects it with Ottawa city, on the Ottawa river. Besides which advantages, it is on the line of the Grand Trunk route. From Prescott may be seen the windmill and the ruined houses, mementoes of the attempt at invasion by Schultz and his band in 1838.

Ogdensburg, New York, the western

Kingston, Canada West.

terminus of the Northern Railway from Lake Champlain, is opposite Prescott.

Maitland, built upon the site of an old French fort, is seven miles above Prescott.

Brockville is yet five miles more, onward. It is one of the best built towns in Canada West.

Gananoque is 32 miles above Brockville.

At **Kingston**, 20 miles yet beyond Gananoque, we leave the St. Lawrence, and approach the waters of Lake Ontario, having previously made our way for an hour through the wonderful labyrinth of the famous "Thousand Isles." See *The Lake of a "Thousand Isles," in the chapter on Canada.* Wolfe's Island, a well-cultivated spot, is opposite Kingston. The city of Kingston, modern as it appears, looks far back for its history, as its advantageous locale did not fail to attract the notice of the early French discoverers. It was once occupied as a small fort called Cataraqui, otherwise known as Frontenac, and was the scene of various sieges and exploits before it passed with all the territory of the Canadas from French to British rule. It was from this point that murderous expeditions were made by the Indians in the olden times against Albany and other English settlements of New York; which in turn sent back here its retributive blows. The present city was founded in 1783. It has now a population of about 16,000. Among its objects of interest, are the fortifications of Fort Henry, on a hill upon the eastern side of the Rideau Canal; two Martello Towers off the town; and other defensive works; the University of Queen's College; the Roman Catholic College of Regiopolis; and the Provincial Penitentiary a little to the west of the city.

As the navigation of the St. Lawrence ends at Kingston, the river boats are exchanged here for others more suited to the lake voyages.

Lake Ontario—American shore.— Let us, before we enter the great waters of Ontario, say a word to the traveller who may prefer to make the voyage along the *American* or lower shore of the lake. From the boundary line 45° the entire *littoral* is in the State of New York.

French Creek comes into the St. Lawrence as we leave it. It was here that General Wilkinson embarked (November 1813) with 7,000 men, with the purpose of descending the river and attacking Montreal. A week subsequently, an engagement took place near Williamsburg, on the Canadian side, when the Americans came off but poorly. General Wilkinson being disappointed in his expectation of reinforcements from Plattsburg, retired to French Mills, and there went into winter-quarters. This place was afterwards named Fort Covington, in memory of General Covington, who fell at the battle of Williamsburg.

Sackett's Harbor, (N. Y.,) lies 20 miles below the mouth of French Creek. It is on the eastern extremity of the lake, on the south side of Black River Bay. This was the naval station of the United States during the English and American War of 1812. In May 1813, Sir George Prevost made a landing with 1,000 troops, but re-embarked without accomplishing any thing. The Navy Yard here is a prominent object as we land.

Oswego, (N. Y.,) is the chief commercial port on the American shore of Ontario. It is very agreeably situated at the mouth of the Oswego river. The Oswego Canal comes in here (38 miles) from Syracuse, and the railway, also, from the same place.

Carthage, the port of the city of Rochester, (N. Y.,) is at the approach to Lake Ontario, of the beautiful Genesee river. (See "Rochester.")

From the mouth of the Genesse to Fort Niagara, a distance of 85 miles, the coast now presents a monotonous and forest-covered level, with a clearing only here and there.

Having now peeped at the American, or southern shore, we will go back to Kingston, and start again on the upper side of the lake, making first for Toronto, 165 miles distant; from Montreal, 333 miles.

Coburg, with a population of about

5,000, is 70 miles from Toronto, and 90 miles from Kingston. It has many and varied manufactories. A railway from Peterboro' (30 miles distant) comes in here. In the vicinage is the *Victoria College*, founded by Act of the Provincial Legislature in 1842.

Port Hope is seven miles above Coburg. From this point, or from Coburg, the journey to Kingston is often charmingly made *overland*, through a beautiful country at the head of the *Bay of Quinté*, a singular arm of the St. Lawrence.

Toronto is the largest and most populous city in Canada West. Sixty years ago the site of the present busy mart was occupied by two Indian families only. In 1793, Governor Simcoe began the settlement under the name of York, changed, when it was incorporated in 1834, to Toronto—meaning, in the Indian tongue, "The place of meeting." The population, in 1817, numbered only 1,200; in 1850, it had reached 25,000; and now, it is, perhaps, 50,000, or upwards.

The Provincial Legislature meets at Toronto and Quebec, alternately, every four years—an arrangement made since the disturbances of 1849, which resulted in the burning of the Parliament Houses at Montreal.

Among the public buildings of Toronto, the traveller will perhaps please himself with a peep at the Catholic Church of St. Michael, the St. James' Cathedral (English), the University of Toronto, the St. Lawrence Hall and Market, the Parliament House, Osgoode Hall, the Post Office, the Court-House, the Exchange, the Mechanics' Institute, Knox's Church, Trinity College, Upper Canada College, the Lunatic Asylum, the Jail, and the Normal and Model Schools. At Toronto, the traveller may, if he pleases, reach Niagara direct, without touching at Hamilton, as we propose to do in our present journey.

Hamilton is among the most beautiful and most prosperous cities of Canada. It aspires, even, to run a race with Toronto, one of the "2.40" nags of the province. Many advantages promise it a brave future. It is

at the head of the western extremity of Lake Ontario, connected with the eastern capitals of the United States, and with Quebec, Montreal, and Toronto, by the Grand Trunk, and the Hamilton and Toronto Railways; and with Lake Huron and the Mississippi States, by the Great Western Railway, which traverses the garden lands of Canada; and, via the Suspension Bridge at Niagara, with the whole railway system of New York. The distance from Toronto to Hamilton, by the steamer, is 45 miles—time, two and a half hours; by railway, 38 miles—time (express), 1 hour 24 minutes. The population of Hamilton, in 1845, was 6,500; at this time it much exceeds 20,000.

From Hamilton to the Falls. Distance, by the Great Western Railway, from Hamilton to the Suspension Bridge, 43 miles—time, 1 hour, 35 minutes. Stations, Ontario, Grimsby, Beamsville, Jordan, St. Catherine's, Thorold, Niagara Falls.

St. Catherine's is the chief point of interest on this part of our route. Its pleasant topography, and, more particularly, its *mineral waters*, is making it a place of great summer resort. Here we leave the reader to establish himself at Niagara, and to see all its marvels, having elsewhere pointed out where he should go, and what should be his *itinéraire* while there. See *Niagara Falls* (New York).

THE GREAT LAKES.

A delightful tour, of a few weeks, may be made, in the heat of the summer, among the natural wonders of the region of the Great Lakes, to Mackinac, the Sault de St. Marie, and the shores of Lake Superior, returning, perhaps, by some one of the lower routes to the Atlantic, from the head waters of the Mississippi.

At Toronto, on Lake Ontario, which may be easily and speedily reached by routes which we shall hereafter travel —from New York, by the Hudson river and Lake Champlain to Montreal, and thence by the Grand Trunk Railway; or by the Central road from Albany to

Buffalo, and by Niagara; or, by Niagara, via the New York and Erie Railway; or, from Portland or Boston, by railroad to Montreal, &c. By Steamboat, daily, from Buffalo, or from Chicago, &c., to Mackinac or *Mackinaw*, as the word is pronounced.

At Toronto, the traveller will take the Collingwood route, by the Ontario, Simcoe, and Huron Railway, 94 miles to Collingwood at the head of Georgian Bay or Manitoulin Lake, the north-east part of Lake Huron. Huron is the third in size of the five great inland seas, which pour their floods into the St. Lawrence. It lies between 43° and 46° 15′ north latitude, having the State of Michigan on the south-south-west, and Canada West upon all other points, excepting where the Straits of Mackinac and the Falls, or Sault de Ste. Marie enter it from Lakes Michigan and Superior, and at its outlet in the St. Clair river. It is divided by the peninsula of Cabot's head, and the Manitouline Islands, the upper portions being the north channel and the Georgian Bay, which we reach at Collingwood. The length of Lake Huron, following its crescent shape, is about 280 miles, and its greatest breadth, not including the Georgian Bay, is 105 miles; its average width is 70 miles. Lake Huron is 352 feet above Lake Ontario, and 600 feet above the level of the sea. The depth is 1,000 feet—greater than that of any other in the grand chain of which it is a link. Off Sagenaw, leads, it is said, have been dropped to a depth of 18,000 feet, which is 12,000 feet below the level of the Atlantic, and yet without finding bottom. The waters here are so pure and clear that objects may be distinctly seen from 50 to 100 feet below the surface. In these noble waters there are said to be more than 3,000 islands.

From Collingwood, the route is by suitable steamers to Mackinac, or the Straits of Mackinac, which are the connecting links between Lake Michigan and Lake Huron. The Island of Mackinac has a circumference of about nine miles, and its shores and vicinage are picturesque and romantic in the highest degree. The Arched Rock facing the water, and rising to the elevation of some 200 feet, makes a bold and striking picture from all points on the lake, and especially as you look through its rude arches from the summit. Robinson's Folly is an attractive bluff on the north shore—years ago a Mr. Robinson, after whom the bluff is named, erected a summer-house upon its crest. Here he passed his days, and oftentimes his nights, despite the cautions of the people about him, until, in an unlucky tempest, he and his eyrie nest were swept away together.

The Cave of Skulls is upon the western shore of the island. Once upon a time, it is said, a party of Sioux Indians were pursued hither by the Ottawas, who imprisoned and destroyed their foes in this cavern, by building fires at its mouth. The traveller, Henry, was one night secreted here, by a friendly Indian, when, to his surprise and horror, the morning light showed that he had been sleeping soundly upon a bed of human bones.

The Needles, another natural wonder of Mackinac, is a bold rock, in form not unlike a light-house. This elevation commands a panorama of the entire island, and a fine view of the crumbling and weed-covered ruins of Fort Holmes. Days of delight may be passed amidst the natural beauties of land and water at Mackinac, made doubly picturesque by the wild frontier life yet found here, and mingled, too, with the still existing homes and presence of the Red men.

Fort Mackinac stands upon a rocky height, 150 feet above the village, which it overlooks. An agency for Indian affairs is established here, which is, from time to time, the resort of deputations and bands of the wild dwellers of the surrounding wilderness. Immense quantities of fish are sent from Mackinac. Steamboats from Detroit, Chicago, and other places, stop here continually.

Sault Ste. Marie. Passing on towards Lake Superior, a voyage of eight pleasant hours, in a steamer, will bring us to the famous Falls of St. Mary, in

Sault St. Marie.

the Strait of St. Mary, which connects the waters of Lake Superior and Lake Huron, and separates Canada West from the upper part of Michigan. The strait extends 63 miles from the south-east extremity of Luke Superior until it reaches Lake Huron. Its course is sometimes narrow, and broken into angry rapids; again, it widens into beautiful lakelets, and winds amid enchanting islands. It is navigable for vessels drawing eight feet of water, up to within a mile of Lake Superior, where the passage is interrupted by the great "Sault" or Falls. The Sault is a series of turbulent rapids, with a total descent of 22 feet in the course of three quarters of a mile.

The exhilarating sport of descending this passage, or of *running the rapids*, as the adventure is called, is that described in the Lake Superior Journal— "Wast thou ever in a gondola in Venice?" is nothing to the question constantly asked nowadays, "Hast thou ever run the St. Mary's Rapids in a birchen canoe?" One who can decide that interesting question in the affirmative, can boast of the most delicious sport ever enjoyed on the water.

The Rapids, bright and sparkling, and white with foam, are beautifully broken and dashed into a thousand eddies by small islands and huge boulders; some rearing their frightful heads above the surface, some bowing themselves under the foaming, rushing currents, seeming like so many sea-monsters forcing their way madly up the rapid stream.

The bark-canoe is beautifully adapted to this "leap;" light as a feather in the hands of its only true master and builder, the Indian, it bounds with every motion of the dancing waters. None but an Indian can pilot our bark down among those furious waters and frightful rocks; but guided by an Ojibwa, a people rocked from their infancy in these birchen canoes, we feel safe as we do on land.

For a perfect enjoyment of this descent, the day should be warm, calm, and clear—so warm that it is a luxury to be out upon these cool waters, so still that not a breath of air can be felt stirring. Though the excitement of the trip never wears off, yet the first venture in the frail boat has a freshness of interest never to be forgotten. One walks up to the "Head," beside the roaring rapids, where the boat is in waiting for the leap, and the never-ceasing roar fills his mind, as well as his ear, with vague sensation of fear; he sees everywhere the ugly-looking boulders in the channel of the river; he sees the waters dashed into foam around them; he has heard of all the accidents that have happened in making the descent, and he steps into the giddy little craft with more of fear than pleasure. But that sensation is of short duration. He feels encouraged by the ease and pleasant coolness of his Indian *voyageurs* and by their dexterity in guiding the canoe.

At first the current is smooth and unbroken, and one looks down through the crystal water at the boulder-covered bottom, and is surprised and delighted at the canoe's rushing into an abyss, gradually subsiding into the quiet flow of a broad river, so that steamboats may run into the very foot of the rapids, and the frail bark-canoe of the adventurous and skilful half-breed dash far up among them in pursuit of the white fish. These rapids are broken up into several different channels, and among them are scattered little islands, such as you see at Niagara, and, like them, bristling with cedars in all possible attitudes.

At this point, on the American side, is the little village of the Sault—an old settlement in the State of Michigan, founded by the Jesuits about two centuries ago. It has evidently seen and felt nothing of the great progress which has been building up cities and states. Here is to be seen the native owner of the soil and the half-breed (a cross of the French and Indian blood); and many other objects of interest.

These rapids are not unlike those of Niagara, excepting that, instead of ending upon the brink of a terrible precipice, they decline with the steady flow of a wide river; and steamers and canoes may fearlessly enter them. They run in different channels, everywhere dodging the numerous little cedar-covered islands in their way. The Sault yields abundant supplies of finny inhabitants; for the excellence of its white fish it is particularly renowned.

The village of the Sault on the Michigan shore, was founded by the Jesuits 200 years ago, but so little progress has it made, that the Aboriginal owner of the soil is still found in possession. Upon the British side of the river, there is an ancient-looking establishment, occupied as an agency of the Hudson's Bay Company.

The St. Mary's Ship Canal, a noble work, now overcomes the obstruction made by these rapids in the passage from Lake Huron to Lake Superior. Heretofore, merchandise from Chicago, Detroit, Buffalo, and other places, had to be discharged and conveyed over a railroad to the upper end of the Sault, and then hauled down to the waters at the opposite extremity; and the locks in this massive canal are, perhaps, the largest in the world.

The Chippewa Hotel is a good house on the American side of the rapids; and Pine's Hotel is a well-kept establishment on the British shore.

Steamers leave the Sault, daily, for all places on Lake Superior, and the neighboring waters.

Lake Superior. We enter Lake Superior after the passage of the *Sault de Ste. Marie*, between two bold promontories, rising to the height of 200 to 300 feet, called Cape Gro and Cape Iroquois.

This grand inland sea is the largest body of fresh water on the globe. Its greatest length is 420 miles, its extreme breadth is 160 miles, and its circuit, 1,750 miles. On its west and northwest shore is Minnesota, on the southern border are Winconsin and Michigan, while British America lies on all other sides. The waters, which are wonderfully transparent, come by more than two hundred streams, from a basin covering an area of 100,000 square miles. The north, and south, and western parts are full of islands, while in the central portions of the lake there are few or none. In the north, these islands are many of them large enough to afford ample shelter for vessels. The picturesque regions of the lake are along the northern shore. In this direction the scenery is of a very bold and striking character. For many miles here there are continuous ranges of cliffs, which reach sometimes an elevation of 1,500 feet; on the south, the banks are

A View of a part of Lake Superior.

Red Sandstone Bluffs, Lake Superior.

low and sandy, except where they are broken by occasional limestone ridges. These ridges rise near the eastern extremity, upon this side, 300 feet, in unique and surprising perpendicular walls and cliffs, broken into the oddest forms, indented with grotesque caverns, and jutting out into ghostly headlands. It is these strange formations which are famous under the name of the " Pictured Rocks." This range is on the east of Point Keweenaw. The rocks have been colored by continual mineral drippings. A similar rocky group lies to the west of the Apostle Islands. It is some hundred feet high, and is broken by numberless arches and caves of the most picturesque character. On the summit of these bluffs, there is everywhere a stunted growth of Alpine trees.

The Porcupine Mountains upon the southern shores of the lake, appear, says a voyager, to be about as extensive (though not so lofty) as the Catskills.

Of the islands of Lake Superior, the largest, which is some 40 miles in length, and from seven to ten broad, is called Royal Isle. Its hills rise to the altitude of 400 feet, with fine bold shores, on the north, and many fine bays on the south. It is, like all this region, a famous fishing-ground. Near the western extremity of the lake, there is a group known as the Apostle Islands. They form a trio of forest-covered heights, adding greatly to the beauty of the landscape around ; on the extreme end of the largest, is the trading post called *La Pointe*, inhabited by Indians and white adventurers. It is a great place of annual rendezvous for the red man and the trader, and a starting point for tramps to the regions of the Mississippi.

The shores of Lake Superior have long been extensively explored for their abundant copper wealth ; and mines have been opened at all points.

Fond du Lac is in Minnesota, on the Saint Louis river, 22 miles from its entrance into Lake Superior. It is accessible by steamboat ; and its wonderfully wild and romantic hills, and rocks, and glens, are well worth a visit from the tourist of the Great Lakes.

We shall come back to this region, when we visit the head waters of the Mississippi, by-and-by.

NEW BRUNSWICK.

THIS Province of Great Britain, lying on the east boundary of the State of Maine, may be reached at St. John's by steamers from Boston (leaving that city every Monday and Thursday, at 9 A.M.; or when the route shall be completed), by the British steamships to Halifax, and thence by railway, 135 miles, to St. John's.

The landscape of New Brunswick is of great variety, and of most pictu-resque beauty; the whole Province (excepting the dozen miles lying directly on the sea) being broken into attractive valleys and hills, which northward assume a very marked, and sometimes, very rugged aspect. Much of its area of 230 miles in length, and 130 in breadth, is covered with magnificent forests, which, as in the neighboring State of Maine, constitute its chief source of industry and wealth.

The hills are nowhere of a very wonderful height, but they often rise in pre-cipitous and sharp acclivities, which give them almost an Alpine aspect; all the more striking in contrast with the peaceful plains and vales which they protect from the tempests of the sea.

Like the neighboring Province of Nova Scotia, New Brunswick so abounds in lakes and rivers, that ready water-access may be had with the help of a short portage, now and then, over its entire area. Thus a canoe may easily be floated from the interior to the Bay of Chaleaux, the Gulf of St. Lawrence, and the ocean on the north, or to the St. John's River, and thence to the Bay of Fundy on the south.

Island of Grand Manan. Bay of Fundy.

The St. John's River is the largest in New Brunswick, and one of the most remarkable and beautiful in America. It rises in the Highlands which separate Maine from Canada, not very far from the sources of the Connecticut. For 150 miles it flows in a north-east di-rection, to the junction of the St.

Francis. Of this part of the river, 100 miles from the *débouchere* of the north-west channel, is called Walloostook River. From the mouth of the St. Francis, the course of the St. John's is irregularly E.S.E. to the Grand Falls ; at which point it makes a descent of from 70 to 80 feet, presenting a splendid picture for the gratification of the tourist. The leap of the Grand Falls past, the river makes its way almost southward for some distance, after which it turns abruptly to the eastward, and so continues its way for 100 miles, passing Fredericton, to the outlet of the Grand Lake, in the southern central part of the Province. From Grand Lake its passage is in a wide channel, due south to Kingston, and thence south-west to St. John's, at its mouth in St. John's Harbor, on the Bay of Fundy.

The entire length of this beautiful river is about 600 miles, and from the Grand Falls to the sea, 225 miles, its course is within the British territory. The boundary line between Maine and the Province lies in the deepest part of the channel of the St. John's, for 75 miles, from a point three miles above the Grand Falls, up to the mouth of the St. Francis River; above the St. Francis the line of the river for 112 miles is entirely in the State of Maine, except-ing a distance of 38 miles through which Canada lies upon the left bank, the right bank only being in Maine. Vessels of 128 tons navigate the St. John's as far up as Fredericton, 80 miles from its mouth, and small steamers go 65 miles beyond, to Woodstock, some-times extending their way to the foot of the Grand Falls. Above the Falls, steamboats pass 40 miles to the mouth of the Madawaska River, and from this point boats and canoes pass *ad libitum*, to the remotest sources. Strenuous efforts have been making by the govern-ment for some years to improve the navigation of the river, and it is hoped that all obstructions will eventually be thoroughly overcome above the Grand Falls. This great river and its affluents are thought to afford 1,300 miles of navi-

gable waters. Very much of the shores of the St. John's is wild forest land. In some parts, the banks rise in grand rocky hills, forming in their lines and interlacings pictures of wonderful de-light.

The chief tributaries of the St. John's, besides the St. Francis and other waters already mentioned, are Aroostook, the Oromocto, and the Alagath, on the west ; and the Salmon, the Naskwaak, the Tobique, the Kennebekasis, and the Washedemoak, from the east.

The coast, and bays, and lakes, and rivers of New Brunswick abound with fish of almost every variety, and in im-mense supplies. The fisheries of the Bay of Fundy are of great value, and employ vast numbers of the population. In the harbor of St. John's alone, there have been at one time two hundred boats, with five hundred men taking salmon, shad, and other fish. Nearly six hundred fishermen have been seen at one period at the Island of Grand Ma-nan ; while at the West Isles, about seven hundred men have been thus em-ployed at one moment ; and so on, at many other of the countless fishing grounds and stations of the New Bruns-wick and the Nova Scotia coasts.

The climate here is healthful, but sub-ject to great extremes of heat and cold ; the mercury rising sometimes to 100° in the day time, and falling to 50° at night.

INTERNAL COMMUNICATION. Besides the steamers and stages which connect the various towns and cities of New Brunswick and Nova Scotia, lines of railway are in active progress, which will unite the two Provinces, and both to the Canadas and the States. A railway from Halifax to the north-ern coast of Nova Scotia, continued by boats, across to the Upper Pro-vince, will unite with the East and North American railway, to a junction with the routes of Canada and the States. Another road is to extend from St. Andrews to Woodstock, and thence to Quebec. The magnetic telegraph already connects New Brunswick, Nova Scotia, and Prince Edward's Island with the States. The connection between

Nova Scotia and Prince Edward's Island, is by a submarine cable, nine miles from Cape Tormentina to Cape Traverse.

St. John's, at the mouth of the St. John's River, is the principal city of New Brunswick, with a population of over 23,000. It is superbly situated upon a bold, rocky peninsula, and is seen very imposingly from the sea. The scenery of the St. John's River is very striking in the passage immediately preceding its entrance into the harbor, and a mile and a half above the city. It makes its impetuous way here in a chain of grand rapids, through rugged gaps, 240 feet wide, and 1,200 feet long. This passage is navigable only during the very brief time of high and equal tides in the harbor and the river; for at low water the river is about 12 feet higher than the harbor, while at high water, the harbor is five feet above the river. It is thus, only, when the waters of the harbor and of the river are on a level, that vessels can pass; and this occurs only during a space of from fifteen to twenty minutes, at each ebb and flow of the tide. Immense quantities of timber are rafted down from the forests of the river, above, to St. John's. It is the entrepôt also of the agricultural and mineral products of a wide region of country.

Fredericton, the capital of New Brunswick, stands upon a flat tongue of land, in a bend of the St. John's River, 80 miles from its mouth. This sandy plain is about three miles long, sometimes reaching a breadth of half a mile. The river, which is navigable up to this point, is here three-quarters of a mile wide. Small steamers ascend 60 miles yet above to Woodstock, and sometimes to the foot of the Great Falls. The population is about 5,000.

The view both up and down the valley is most interesting—to the north an uncleared range of highlands, with detached cones and broken hills thrown out in bold relief upon the landscape. Villas enclosed in the woods, and farms upon the clearings, are the chief objects it presents ; while to the south the river is seen widening, like a silver cord, through the dark woodlands, until it disappears among the islands in the distance.

St. Andrew's, with a population of about 8,000, is at the north-east extremity of Passamaquoddy Bay, three miles from the shores of the United States, near Eastport, in Maine, and 60 miles from St. John's. A railway will connect St. Andrew's with Woodstock, 80 miles distant, and with other routes to the Canadas.

NOVA SCOTIA.

NOVA SCOTIA, the ancient Acadia, including the Island of Cape Breton and Sable Island, lies south-east of New Brunswick, from which it is separated by the Bay of Fundy, except only at the narrow Isthmus of Chignecto. It may be reached at Halifax, its capital, by the British steamers from New York and Boston. The railways now in progress within its limits will soon more conveniently unite it to the cities of the Canadas and the United States. The area of the Province is 18,746 square miles, including the 3,000 of Cape Breton, and the 69 of Sable Island. The shores are bleak, and often very rugged. The interior is diversified with hills and valleys, though not of very bold character, as the highest land is but 810 feet above the sea. The numerous lakes cover a great part of the Province. The soil is not remarkably productive, excepting the fertile lands upon the rivers and bays. The richest portion is that bordering the Bay of Fundy. The most thickly-settled and productive region is that about the Minas basin, while the neighborhood of Halifax, on the southern coast, is the most im-

3

portant. Much of Nova Scotia is, like New Brunswick, forest land, which makes the lumber trade very large and productive. This, however, is not the only industry of the people, as the abundant irrigation of the valley lands makes agriculture, especially in the raising of grass, very remunerative. Upon the coast, too, many people are occupied in the extensive fishing trade, which has been prosecuted here more actively than upon any of the British American shores, excepting only that of Newfoundland.

THE COAST OF NOVA SCOTIA.—The greatest length of Nova Scotia is 280 miles, and the greatest breadth 120 miles. The south-east coast, in a distance of 110 miles only from Cape Canso to Halifax, has no less than 12 ports capacious enough to receive ships of the line, and 14 deep enough for merchantmen. A belt of rugged broken land, of the average height of 500 feet, formed of granite and primary rock, extends along all the Atlantic shore, from Cape Canso to Cape Sable. This belt varies in breadth from 20 to 60 miles. Similar formations are found over two-thirds of the whole Province. From Briar's Island, off Digby Neck, 130 miles to Capes Split and Blomidon, along the northern coast on the Bay of Fundy, there is a ridge of fir-covered frowning precipices of trap rock, which overhang the waves at an elevation of from 100 to 600 feet. These magnificent cliffs are picturesque and grand in the extreme. They are, too (which is something in this utilitarian age), not only ornamental but useful, for they serve to protect the interior from the terrible fogs of the bay.

THE RIVERS AND LAKES AND BAYS OF NOVA SCOTIA.—The lakes here, though generally small, are almost countless in number, covering the whole peninsula as with a net-work of smiling waters. In some instances, no less than a hundred are grouped within a space of 20 square miles! Lake Rosignol, the largest of the region, is 30 miles long. It is near the western end of the peninsula. Great Lake comes next, then College Lake, eastward. Mines Bay on the north coast, the eastern arm of the Bay of Fundy, penetrating 60 miles inland, is very remarkable for the tremendous tides which rush in here, sometimes to the height of 60 to 70 feet, while they do not reach more than from 6 to 9 feet in the harbor of Halifax, directly opposite; these are the spring-tides. They form what is called the *bore*. The Bays of St. Mary's, the Gut of Canso, Townsend Bay, George Bay, and Chedabucto Bay, in the western part of the Province, and St. Margaret's and Mahone Bays on the south, are all large and most interesting waters.

The Annapolis River flows into the Bay of Fundy, 100 miles from the Garden of Acadia. Besides this principal river there are many others navigable for a greater or less distance from their mouths, as the Shubenacadie, which, by the help of a canal, connects Cobequid Bay, from the Bay of Fundy on the north side of the peninsula, with Halifax Harbor on the south; the Musket and the Clyde in the south-west extremity of the Province, the Mersey, the Musquodoboit, and the St. Mary's. Indeed, rivers pour their waters into all the many bays and harbors which so thickly stud the whole line of these remarkable coasts.

Halifax, the capital of Nova Scotia, is upon the south coast of the peninsula, on the declivity of a hill, about 250 feet high, rising from one of the finest harbors on the continent. The streets are generally broad, and for the most part macadamized. Viewed from the water, or from the opposite shore, the city is prepossessing and animated. In front, the town is lined with wharves, which, from the number of vessels constantly loading and discharging, always exhibit a spectacle of great commercial activity. Warehouses rise over the wharves, or tower aloft in different parts of the town, and dwelling-houses and public buildings rear their heads over each

Halifax, Nova Scotia.

other, as they stretch along and up the sides of the hill. The spires of the different churches, the building above the town, in which the town-clock is fixed, a rotunda-built church, the signal-posts on Citadel Hill, the different batteries, the variety of style in which the houses are built, some of which are painted white, some blue, and some red; rows of trees showing themselves in different parts of the town; the ships moored opposite the dockyard, with the establishments and tall shears of the latter; the merchant vessels under sail, at anchor, or along the wharves; the wooded and rocky scenery of the background, with the islands and the small town of Dartmouth on the east shore—are all objects most agreeable to see.

Of the public buildings, the chief is a handsome edifice of stone, called the Province Building, 140 feet long by 70 broad, and ornamented with a colonnade of the Ionic order. It comprises chambers for the Council and Legislative Assembly, the Supreme Court, various government offices, and the public library of the city. The Government House, in the southern part of the town, is a solid, but gloomy-looking structure, near which is the residence of the military commandant. The Admiral's residence, on the north side of the town, is a plain building of stone. The Dockyard, which covers 14 acres, and forms the chief depot of naval stores in the British North American colonies, is reported to be the finest in the world, if the works of a similar kind in England be excepted. The north and south barracks are capable of accommodating three regiments. There is also a Military Hospital, erected by the late Duke of Kent. Dalhousie College is a handsome edifice of freestone. The Church of England and the Presbyterians have each churches; and there are beside a Roman Catholic chapel, and several churches belonging to different religious denominations. There are two private banking companies in the town, and a chamber of commerce. The hotels and boarding-houses are not of the highest order. The inhabitants of Halifax are intelligent and social, and travellers will remark a tone of society here more decidedly English than in most of the other colonial cities.

The harbor opposite the town is more than a mile wide, and has, at medium tides, a depth of 12 fathoms. About a mile above the upper end of the town it narrows to one-fourth of a mile, and then expands into Bedford Basin, which has a surface of 10 square miles, and is completely land-locked. On an island opposite the town are some strong mounted batteries. The harbor is also defended by some other minor fortifications.

Halifax, ever since its settlement in 1749, has been the seat of a profitable fishery. Its trade, which is in a very prosperous condition, is principally with the West Indies and other British colonies, with the United States, and the mother country. It is also the chief rendezvous and naval depot for the British navy on the North American station. The British Government having made Halifax one of the stopping-places of the Cunard line of steamers, in their trips either way across the Atlantic, has added greatly to its importance as a maritime city, as well as advanced its commercial prosperity.

The population of Halifax is about 30,000.

County Map of the
SOUTHERN

NEW ENGLAND STATES,

Showing the Railroads
and their connection with the
CITIES OF

NEW-YORK, BOSTON & ALBA[NY]

WITH DISTANCES.

Scale of Miles.

0   5 10   20   30

Longitude East   5   from Washington.   6

ATLANTIC OCEAN

MASSACHUSETTS BAY

CAPE COD BAY

PLYMOUTH

PROVIDENCE

NEW LONDON

WASHINGTON

KENT

BARNSTABLE

MARTHA'S VINEYARD

Montauk P.

Block I.
to Rhode I.

Fishers I.
to New York.

Sag Harbour

THE UNITED STATES.

———•♦•———

The grand territory of the United States, through which we propose to travel in our present volume, occupies no meaner area than that of 2,936,166 square miles, scarcely less than that of the continent of Europe. In form, it is nearly a parallelogram, with an average length of 2,400 miles, from east to west, and a mean breadth, from north to south, of 1,300 miles. Its extreme length and breadth are, respectively, 2,700 and 1,600 miles; reaching from the Atlantic on the east, to the Pacific on the west; from British America on the north, to the Gulf of Mexico and the Mexican Republic on the south. Its present division is into thirty-three States and seven Territories, including the District of Columbia. The States have been popularly grouped into 4 classes, according to their geographical position; as the Eastern Group, or "New England," embracing Maine, New Hampshire, Vermont, Massachusetts, Rhode Island, and Connecticut ; the "Middle" group of New York, New Jersey, Pennsylvania, Delaware, and Maryland ; the "Southern States," Virginia, North Carolina, South Carolina, Georgia, Florida, Alabama, Mississippi, Louisiana, and Texas ; and the "Western States" of Tennessee, Kentucky, Ohio, Indiana, Illinois, Michigan, Iowa, Wisconsin, Missouri, Arkansas, California, Oregon, and Minnesota.

All the Territories—Washington, New Mexico, Utah, the Indian, Kansas, and Nebraska, are included in this division of the country.

POPULATION OF THE UNITED STATES.

*The District of Columbia, (D. C.,†) 1850............................... 51,687

THE EASTERN OR NEW ENGLAND STATES.

Connecticut, (Conn.,) 1850..........	370,792	New Hampshire, (N. H.,) 1850......	317,976
Rhode Island, (R. I.,) 1850..........	147,545	Vermont, (Vt.,)....................	314,120
Massachusetts, (Mass.,) 1855........	1,132,369	Maine, (Me.,) 1850	583,169

THE MIDDLE STATES.

New York, (N. Y.,) 1855...........	3,466,212	Pennsylvania, (Pa.,) 1850...........	2,311,786
New Jersey, (N. J.,) 1850	489,555	Delaware, (Del.,) 1850..............	91,532

THE SOUTHERN STATES.

*Maryland, (Md.,) 1850	583,034	*Florida, (Fa.,) 1855	110,823
*Virginia, (Va.,) 1850.	1,421,661	*Alabama, (Ala.,) 1855	841,704
*North Carolina, (N. C.,) 1850.......	869,039	*Louisiana, (La.,) 1859	646,971
*South Carolina, (S. C.,) 1850.......	668,507	*Texas, (Tex.,) 1850................	212,592
*Georgia, (Ga.,) 1859...............	1,014,418	*Mississippi, (Miss.,) 1850	606,526

* Slave States.   † Abbreviations used in the address of letters, etc.

General Remarks—Rivers and Lakes.

THE WESTERN STATES.

*Arkansas, (Ark.,) 1858	331,213	Illinois, (Ill.,) 1855	1,306,576
*Tennessee, (Tenn.,) 1850	1,002,717	Michigan, (Mich.,) 1854	511,672
*Kentucky, (Ky.,) 1850	982,405	Wisconsin, (Wis.,) 1855	552,451
Ohio, (O.,) 1850	1,980,329	Iowa, (Io.,) 1859	633,549
Indiana, (Ia.,) 1850	988,416	*Missouri, (Mo.,) 1850	682,044
Minnesota, (Min.,) 1858	150,042	Oregon, (Or.,) 1857	43,000
	California, (Cal.,) 1856	507,067	

TERRITORIES.

New Mexico, (N. M. Ty.,)	61,547	Utah, (Ut. Ty.,)	11,380
Washington, (Was. Ty.,) 1857	10,000	Nebraska, (Na. Ty.,) 1856	10,716
Kansas, (Ka. Ty.,) 1858	75,000	Indian Territory, (Ind. Ty.,)	187,171
	Total	26,279,293	

MAINE.

MAINE is the extreme eastern portion of New England, and the border State of the Union in that direction, with the British province of New Brunswick on the north and north-east, and Canada on the north-west. It has three distinct topographical aspects—in the comparatively level, and somewhat sandy and marshy character of the southern portion, lying back 20 miles from the Atlantic coasts—in the pleasant hill and valley features of the interior, and in the rugged, mountainous, and wilderness regions of the north.

A great portion of the State is yet covered by dense forests, the utilization of which is the chief occupation and support of its inhabitants. The most fertile lands lie in the central southern regions, between the Penobscot river on the east, and the Kennebec on the west, and in the valley borders of other waters. The mountain ranges are often very bold and imposing—one summit, that of Katahdin, having an elevation of 5,385 feet above the level of the sea. The lakes are very numerous, sometimes of great extent, and often very beautiful, all over the State; and more especially among the mountains in the north. Indeed, it is estimated, that one tenth part of the whole area of Maine is covered by water. The rivers are numerous and large, and present everywhere scenes of great and varied beauty. The Atlantic coast, which occupies the whole southern line of the State, is the finest in the Union, in its remarkably bold, rocky character, and in its beautiful harbors, bays, islands, and beaches. The sea-islands of Maine are over 400 in number; and many of them are very large and covered by fertile and inhabited lands. The climate, though marked by extremes, both of heat and cold, is yet everywhere most healthful; and its rigor is much modified by the proximity of the ocean.

The Mountains and Lakes. The most interesting route for the tourist here, is perhaps a journey through the hills, lakes, and forests of the north; but we warn him, beforehand, that it will not be one of ease. Rugged roads and scant physical comforts will not be his most severe trial: for, in many places, he will not find road or inn at all, but must trudge along painfully on foot, or by rude skiff over the lakes, and trust to his rifle and his rod to supply his larder. In these wildest regions the exploration may be made with great satisfaction by a party well provided with all needed tent equipage, and with all the paraphernalia of the chase; for deer, and the moose, and the wild fowl are abundant in the woods; and the finest fish may be freely taken in the water. Still he may traverse most of the mountain lands and lakes by the roads and paths of the lumbermen, who have invaded all the region; and he may bivouac as comfortably as should content an orthodox forester, in these humble shanties.

The mountains of Maine are broken and distinct peaks. A range, which seems to be an irregular continuation of the White Hills of New Hampshire, extends along the western side of the State for many miles; and, verging towards the north-east, terminates in Mars Hill. This chain divides the waters which flow north into the St. John's river from those which pass southward to the Atlantic. Many beautiful lakes lie within this territory.

Mount Katahdin, Maine.

Mount Katahdin, with its peaks 5,385 feet above the sea, is the loftiest summit in the State, and is the *ultima thule*, at present, of general travel in this direction. The ordinary access is in stages from Bangor over the Aroostook Road, starting in tolerable coaches on a tolerable road, and changing always, in both, from bad to worse. A pleasant route *for the adventurer* is down the West Branch of the Penobscot, in a canoe, from Moosehead Lake. Guides and birches, as the boats are called, may be procured at the foot of Moosehead, or at the Kineo House, near the centre of the lake. By this approach Katahdin is seen in much finer outlines than from the eastward.

Sugar Loaf Mountain, upon the Seboois river, north-east of Mount Katahdin, is nearly 2,000 feet high, and from its summit a magnificent view is commanded, which embraces some fifty mountain peaks, and nearly a score of wonderful lakes. Then there are Bigelow, Saddleback, Squaw, Bald, Gilead, the Speckled Mountain, the Blue Mountain, and other heights more or less noble, amidst which are brooks and lakelets and waterfalls of most romantic character.

Moosehead Lake, the largest in Maine, is among the northern hills. It is 35 miles long, and, at one point, is 10 miles in breadth, though near the centre there is a pass not over a mile across. Its waters are deep, and furnish ample occupation to the angler, in their stores of trout and other fish. This lake may be traversed in the steamboats employed in towing lumber to the Kennebec. A summer hotel occupies a very picturesque site upon the shore. The Kineo House, midway, is the usual stopping-place. There are numerous islands on the Moosehead Lake, some of which are of great interest. On the west side Mount Kineo overhangs the water, at an elevation of 600 feet. Its summit reveals a picture of forest beauty well worth the climbing to see. Moosehead is 15 miles north of the village of Monson, and 60 north-west of Bangor. The roads thither, lying through forest land, are necessarily somewhat rough and lonely. This lake is the source of the great Kennebec river, by whose channels its waters reach the sea.

Lake Umbagog lies partly in Maine and yet more in New Hampshire. Its length is about 12 miles, and its breadth varies from 1 to 5 miles. The outlet of Umbagog and the Margallaway river form the Androscoggin.

Androscoggin and Moosetocknoguntic Lakes are in the vicinity of Umbagog.

Sebago Pond, a beautiful lake 12 miles long and from 7 to 8 miles broad, is about 20 miles from Portland, on a route thence to Conway and the White Mountains. It is connected with Portland by a canal.

The Penobscot, the largest and most beautiful of the rivers of Maine, may be reached daily from Boston and Portland, by steamer, as far up as Bangor, and also by railway from Portland to Bangor. It is formed by two branches, the east and the west, which unite near the centre of the State, and flow in a general south-west course to Bangor, 60 miles from the sea, and at the head of navigation. Large vessels can ascend to Bangor, and small steamboats navigate the river yet above. At Bangor the tide rises to the great height of 17 feet, an elevation which is supposed to be produced by the wedge-shaped form of the bay, and by the current from the Gulf Stream. The length of the Penobscot, from the junction of the east and west branch, is 135 miles, or measuring from the source of the west branch, it is 300 miles; though, as far as the tourist is concerned, it is only 60 miles—being that portion between Bangor and the ocean. This part, then, the Penobscot proper, ranks, in its pictorial attractions, among the finest river scenery of the United States. In all its course

there are continual points of great beauty, and very often the shores rise in striking and even grand lines and proportions. We have met tourists who have been hardly less impressed with the landscape of this fine river than with that of the Hudson even, though we do not admit such a comparison.

Bangor, at the head of tide water and of navigation on the Penobscot river, 60 miles from its mouth, is one of the largest cities of Maine, having a population of more than 20,000. Steamboats connect it daily with Portland and Boston; and it is reached also by the Androscoggin and Kennebec, and Penobscot and Kennebec railways, via Waterville, on the Kennebec. The distance from Bangor to Portland, by railway, is 138 miles. Bangor is connected with Old Town (12 miles), by railway, and another road is contemplated to Lincoln, 50 miles up the Penobscot valley. The Bangor Theological Seminary, founded 1816, occupies a fine site in the higher portion of the city. The chief hotel is the Bangor House. The "speciality" of Bangor is lumber, of which it is the greatest depot in the world. All the vast country above, drained by the Penobscot and its affluents, is covered with dense forests of pine, and hemlock, and spruce, and cedar, from which immense quantities of lumber are continually cut and sent from the marvellous saw-mills, down the river to market at Bangor. During the eight or nine months of the year through which the navigation of the river is open, some 2,000 vessels are employed in the transportation of this freight. The whole industry of Bangor is not, however, in the lumber line, as she is also engaged in ship building, and has a large coasting trade, and a considerable foreign commerce.

Belfast and Castine are some 30 miles below Bangor, where the Penobscot enters its name-sake Bay. Belfast, on the west, and Castine on the east shore, are nine miles apart. They are both small ship-building and fishing towns.

The Kennebec River is in the western part of the State, extending from

Moosehead Lake, 150 miles to the sea. It makes a descent in its passage of 1,000 feet, thus affording a great and valuable water-power. The scenery of the Kennebec, though pleasant, is far less striking than that of the Penobscot. Its shores are thickly lined with towns and villages, among which is Augusta, the capital of the State.

Augusta is at the head of sloop navigation on the Kennebec, 43 miles from its mouth. It is 60 miles N.N.E. of Portland by railway, and 69 S.W. of Bangor. Steamboats run hence to Portland and Boston, calling at the river landings. The city is chiefly upon the right bank of the river, which is crossed here by a bridge 520 feet long; and a quarter of a mile above, by a railroad bridge, 900 feet in length. The private residences, and some of the hotels, are upon a terrace, a short distance west of the river, while the business parts of the town lie along shore. The State House is an elegant structure of white granite. Its site, in the southern part of the city, is lofty and very picturesque; in front is a large and well-cared-for park. The United States Arsenal, surrounded with extensive and elegant grounds, is upon the east side of the river. Here, too, is the Hospital for the Insane, built upon a commanding and most beautiful eminence. The principal hotels here, are the Stanley House, the Augusta House, and the Mansion House. Augusta is upon the railway route from Portland to Bangor. Population, 10,000.

Hallowell is a pretty village, two miles below Augusta, on the river, and on the line of the Kennebec and Portland Railway.

Gardiner on the Kennebec, at the mouth of the Cobessecontee River. This point is the head of ship navigation on the Kennebec. The city is seven miles below Augusta, and 53 miles from Portland by the Kennebec and Portland Railway.

Waterville is on the Kennebec, at the Ticonic Falls, and at the northern terminus of the Androscoggin and Kennebec Railway, connecting with the Kennebec and Penobscot line. It is the seat of Waterville College, a prosperous establishment, controlled by the Baptists.

Bath, a flourishing city of over 12,000 people, is on the Kennebec, 12 miles from the sea; 30 miles south of Augusta; and 36 north-east of Portland. It is the terminus of a branch road from Brunswick, on the Kennebec and Portland Railway; it is to be united at Lewiston with the Androscoggin and Kennebec route from Portland to Bangor.

The Androscoggin River is a fine stream, flowing from Lake Umbagog, partly in New Hampshire, but chiefly through the south-western corner of Maine, into the Kennebec, 20 miles from the ocean.

Brunswick, on the Androscoggin, is 27 miles from Portland by railway. It is the seat of Bowdoin College, which is beautifully located on a high terrace, near the edge of the village. This popular institution was founded in 1802. The Medical School of Maine, which is connected with Bowdoin College, has a very valuable library, and anatomical cabinet. The Androscoggin here falls 50 feet within the reach of half a mile.

Mount Desert Island. A summer trip to Mount Desert Island has of late years been a pleasant treat to American landscape painters, and a visit thither might be equally grateful to the general tourist. The vigorous and varied rock-bound coast of New England can be nowhere seen to greater advantage. Mount Desert Island is an out-of-the-way nook of beauty in Frenchman's Bay, east of the mouth of the Penobscot River. It is 40 miles from Bangor, and may be reached from Boston by boat, via Rockville, and thence by another steamer, on to Bucksport (on the Penobscot), and thence by stage via Ellsworth, or from Castine on the Penobscot Bay, hard by. If the visitor here cannot sketch the bold, rocky cliffs, he can beguile the fish to his heart's content.

Eastport, upon the waters of Passamaquoddy Bay, at the extreme eastern point of the territory of the

3*

Mount Desert Island, Maine.

United States, is well deserving of a visit from the tourist in quest of the beautiful in nature ; for more charming scenes on land and on sea, than are here, may rarely be found.

The traveller may see Eastport and its vicinage and then go home, if he pleases ; for it is the jumping-off place —the veritable Land-End—the latitude and longitude beyond which the stars and stripes give place to the red cross of England.

Eastport is 234 miles N.E. of Portland, and is reached thence and from Boston by regular steamboat communication. Steamboats run also to Calais and places *en route*, 30 miles above, at the head of navigation on the St. Croix River. The town is charmingly built on Moose Island, and is connected to the mainland of Perry by a bridge; and by ferries with Pembroke, Lubec, and the adjoining British Islands. It is not a very ponderous place, the population of the township scarcely exceeding 5,000. Fort Sullivan is its shield and buckler against any possible foes from without.

The *Passamaquoddy Bay* extends inland some 15 miles, and is, perhaps, 10 miles in breadth. Its shores are wonderfully irregular and picturesque, and the many islands which stud its deep waters, help much in the composition of pictures to be enjoyed and remembered.

TO PORTLAND, MAINE.

From Boston, 107 miles by the Eastern Railway, via Lynn, Salem, Newburyport, Portsmouth, N. H., &c. ; or by the Boston and Maine route, 111 miles through Reading, Lawrence, Andover, Haverhill, Exeter, Dover, &c. ; or by steamer daily.

From Montreal, by the Grand Trunk Railway.

Portland, the commercial metropolis of Maine, is handsomely situated on a peninsula, occupying the ridge and side of a high point of land, in the S.W. extremity of Casco Bay, and on approaching it from the ocean, is seen to great advantage. The harbor is one of the best on the Atlantic coast, the anchorage being protected on every side by

Portland.

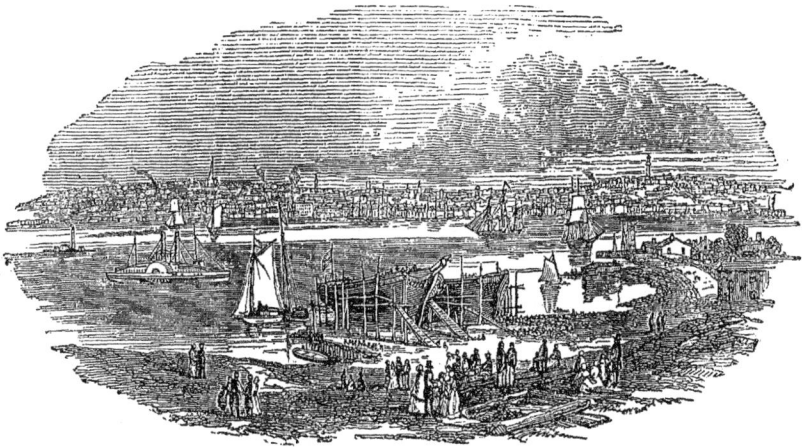

Portland, Maine.

land, whilst the water is deep, and communication with the ocean direct and convenient. It is defended by Forts Preble and Scammel. On the highest point of the peninsula is an observatory 70 feet in height, commanding a fine view of the city, harbor, and islands in the bay. The misty forms of the White Mountains, 60 miles distant, are discernible in clear weather.

This city is elegantly built, and the streets are beautifully shaded and embellished with trees; and so profusely, that there are said to be here no less than 3,000 of these rural delights. Congress street, the main highway, follows the ridge of the peninsula through its entire extent. Among the public buildings of Portland, the City Hall, the Court House, and some of the churches, are worthy of particular attention. The Society of Natural History possesses a fine cabinet, containing specimens of the ornithology of the State, more than 4,000 species of shells, and a rich collection of mineralogical and geological examples, and of fishes and reptiles. The Athenæum has a library of 8,000 volumes, and the Mercantile Library possesses, also, many valuable books. The Portland Sacred Music Society is an interesting Association here.

The long line of the Grand Trunk Railway connects the city with Montreal, in Canada, and thence with all the region of the St. Lawrence River. Two lines of railway unite it with Boston and the western cities, and with the interior of Maine, at Augusta on the Kennebec, and at Bangor, on the Penobscot Rivers.

Hotels. The American, the Elm, and the United States.

The population of Portland is about 30,000.

THE GRAND TRUNK RAILWAY, ROUTE FROM PORTLAND, NORTH. This great thoroughfare connects the navigable waters of Portland harbor with the great commercial capital of Canada. Its route passes through a fertile and productive country, generally under fine cultivation, the streams in its vicinity abounding in water privileges of the first importance. From Portland it passes onward to the valley of Royal's River, and follows up the valley of the Little Androscoggin. It strikes and crosses that river at Mechanic Falls, 43 miles from Portland, at which place the Buckfield Branch Railroad connects with it. Pursuing its course upward, it passes in the vicinity of the "Mills" on its way to Paris Cape, in the neighborhood of Norway and Paris, drawing in upon it the travel and business of that rich and populous region. Still following up the valley of the Little Androscoggin, passing on the way two import-

Route of the Grand Trunk Railway.

ant falls, it reaches Bryant's Pond, the source of that river. This point is 15 miles from Rumford Falls, on the Great Androscoggin, one of the greatest and most available water-powers in the State. Passing hence into the valley of Alder stream, the route strikes the Great Androscoggin, near Bethel, a distance of 70 miles from Portland. Crossing that stream, it follows up its picturesque and romantic valley, bordered by the highest mountains in New England, till, in its course of about 20 miles from Bethel, it reaches Gorham in New Hampshire, distant from the base of Mount Washington a few miles only. From this point that celebrated shrine may be approached and ascended with more ease, in a shorter distance, and less time, than from any other accessible quarter in the vicinity of the White Hills. (See routes to White Mountains.) This point also is only five miles distant from Berlin Falls, the greatest waterfall in New England, where the waters of the Great Androscoggin, larger in volume than the waters of the Connecticut, descend nearly 200 feet in a distance of about two miles. From the valley of the Androscoggin the road passes into the valley of the Connecticut, reaching the banks of that river at North Stratford in New Hampshire.

Following up this rich and highly productive valley about 35 miles, the road reaches the parallel of 45° N. Lat., at the boundary between the United States and Canada; continuing thence to Quebec, and up the St. Lawrence via Montreal, to Toronto on Lake Ontario, where it connects with other routes for Lake Superior and all parts of the great West.

Lewiston is a flourishing manufacturing village, containing about 7,000 inhabitants, situated upon the Androscoggin River between Portland and Waterville on the Androscoggin and Kennebec Railway, 33 miles north of Portland. The waterfall here is one of exceeding beauty. The entire volume of the Androscoggin is precipitated some 50 feet over a broken ledge, forming in its fall a splendid specimen of natural scenery. The river immediately below the fall, subsides into almost a uniform tranquillity, and moves slowly and gracefully along its course, in strange though pleasing contrast with its wild and turbid appearance at and above the cataract. The Androscoggin and Kennebec road communicates with the Grand Trunk Railway at Danville, six miles below Lewiston, and with the Androscoggin road at Leeds, 11 miles above.

MASSACHUSETTS.

The landscape here is of changeful character, and often strikingly beautiful, embracing not a few of the most famous scenes in the Union. In the southeastern part of the State the surface is flat and sandy, though the sea coast is, in many places, very bold, and charmingly varied with fine pictures of rocky bluff and cliff. It abounds in admirable summer houses, where the lovers of sea-breezes and bathing may find every means and appliance for comfort and pleasure.

In the eastern and central regions, the physical aspect of the country, though agreeably diversified, is eclipsed in attraction by the lavish art-adornments of crowding city and village, and happy homesteads, nowhere so abundant and so interesting as here.

The Green Mountains traverse the western portion of Massachusetts in two ridges, lying some 25 miles apart, with picturesque valley lands between. Here are the favorite summer resorts of Berkshire, and other parts of the Housatonic region. Saddle Mountain, 3,505 feet, is a spur of the most western of the two ridges we have mentioned, known as the Taconic or the Taughkannic hills.

OLD COLONY R.R.

WITH PART OF THE

FALL RIVER ROUTE

TO

NEW YORK.

MASSACHUSETTS BAY

M A S S A C H U S E T T S

...CHUSETTS

M O U T H

Gravel
...Cott
...mer
...Brewster
Jer.Brewster
...E.
Boston Light
...lerton
...ket
...L.

Cohasset
...e R.R.
L R E R K
...O

Up. P... ns
Scituate
Scituate Har.
Scituate Har.V.
Liberty Plains
S. Scituate
N.Marshfield
...ith
Jacob's Pd V.
E.Marshfield
East
...ington
Hanover
South

Abington
Corners
S.Marshfield
Oldham E.
Pembroke
Green Harbor
...mittacus
...uit
Furnace P.
Sandy Pds.
...chester
Mattapoisett
...car V.
Bryantville
Jones River P.
Duxbury
Clark's I.
RAIL
...28
Monponset Pond
King-ston 33
Plymouth Har.
B
R
Bridg
30
Jones R.
Halifax
N.Village
ROAD
Plympton
PLYMOUTH
37 m. fr. Boston
Eddyville
Eel River V.
Carver Green
Billington P.
Middleboro'Cen.
G.South P.

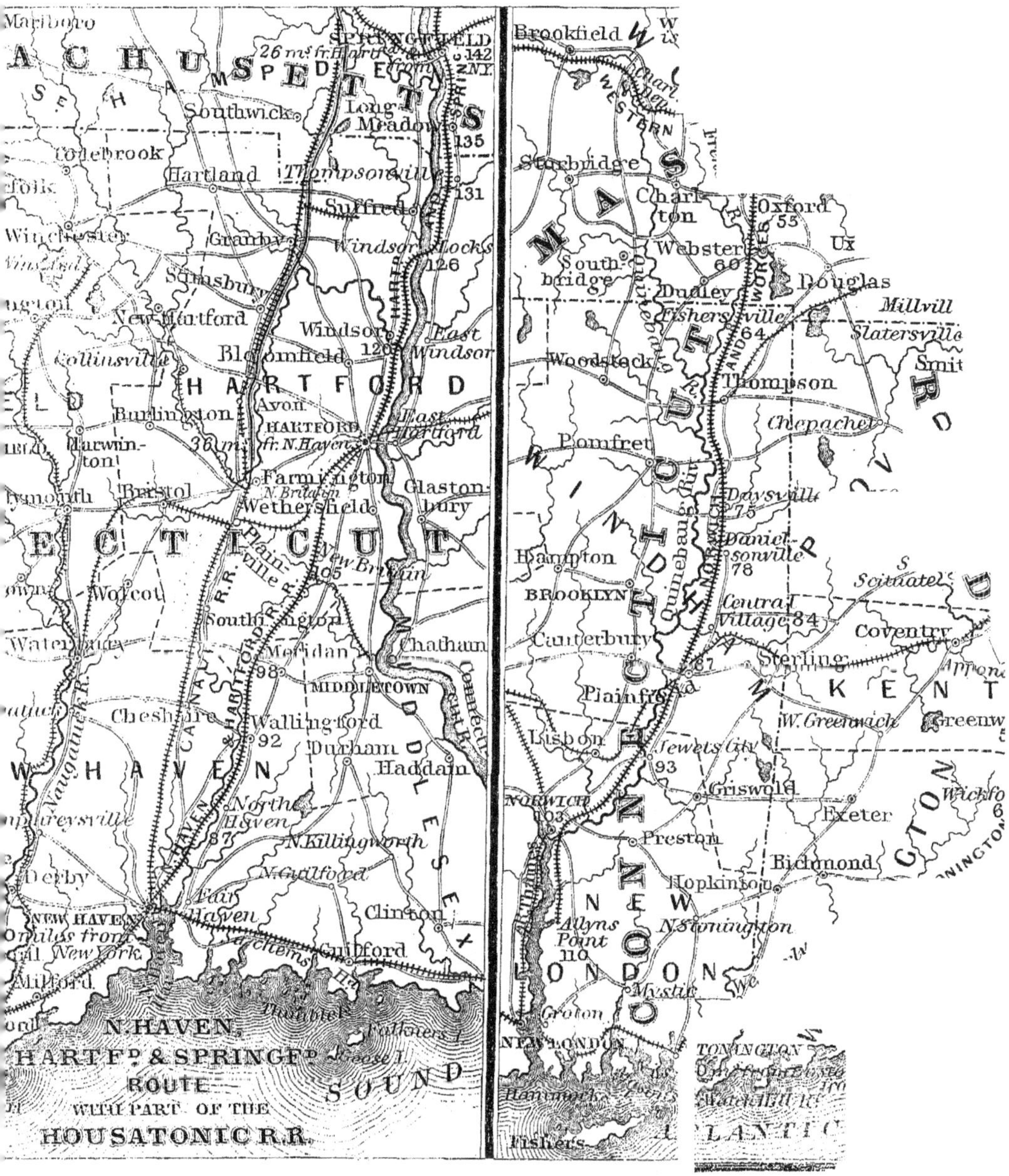

A map with the following visible text labels:

MASSACHUSETTS (spanned as "A C H U S E D T T S")

Marlboro · Southwick · Long Meadow · SPRINGFIELD 142 · Brookfield · W. U. · Codebrook · folk · Hartland · Thompsonville · 135 · Suffield · 131 · Sturbridge · Charlton · Oxford 55 · Winchester · Winsted · Granby · Windsor Locks 126 · Southbridge · Webster 60 · Ux · Simsbury · Douglas · Dudley · Fishersville 64 · Millvill · New Hartford · Windsor 120 · East Windsor · Woodstock · Thompson · Slatersville · Collinsville · BLOOMFIELD · Smit · Burlington · Avon · HARTFORD · East Hartford · Pomfret · Chepachet · Harwinton · 30 m N. Haven · N. Bridge · Farmington · Glastonbury · Daysville 75 · Plymouth · Bristol · Wethersfield · Danielsonville 78 · S. Scituate · Plainville R.R. · New Britain 105 · Hampton · BROOKLYN · Central Village 84 · Coventry · Wolcott · Southington · Canterbury · Sterling · Appon · Waterbury · Meriden 98 · Chatham · Plainfield 87 · W. Greenwich · Greenw · Cheshire · MIDDLETOWN · Wallingford 92 · Durham · Haddam · Lisbon · Jewets City · Wickfo 6 · Naugatuc · North Haven 87 · Guilford · Norwich 103 · Griswold · Exeter · Humpreysville · N. Killingworth · Preston · Richmond · Hopkinton · Derby · N. Guilford · Fair Haven · Clinton · Allyns Point 110 · N. Stonington · NEW HAVEN · Milford · New York · Guilford · Groton · Mystic · NEW LONDON · STONINGTON · Watch Hill R. · Hammock · Fishers

CONNECTICUT · **HARTFORD** · **NEW HAVEN** · **MIDDLESEX** · **WINDHAM** · **TOLLAND** · **NEW LONDON** · **KENT**

SOUND · ATLANTIC

N. HAVEN HARTF'D. & SPRINGF'D. ROUTE WITH PART OF THE HOUSATONIC R.R.

Mount Washington, another fine peak of this line, has an altitude of 2,624 feet. It rises in the extreme southern corner of the State, while Saddle Mountain stands as an outpost in the north-west angle. The more eastern of the two hill-ranges here is called the Hoosic Ridge. Noble isolated mountain peaks overlook the winding waters and valleys of the Connecticut—some of them, though not of remarkable altitude, commanding scenes of wondrous interest, as Mount Holyoke and Mount Tom, near Northampton. North of the middle of the State is the Wachusett Mountain, with an elevation of 2,018 feet.

On Hudson's Brook, in Adams, there is found a remarkable natural bridge, 50 feet high, spanning a limestone ravine 500 feet in length. In New Marlborough the tourist will see a singular rock poised with such marvellous art that a finger can move it; and on Farmington River, in Sandisfield, he will delight himself with the precipices, 300 feet high, known as the Hanging Mountain.

Massachusetts has some valuable mineral springs, though none of them are places of general resort. In Hopkinton, mineral waters impregnated with carbonic acid and carbonates of iron and lime; in Winchenden, a chalybeate spring, and one in Shutesbury, containing muriate of lime. But we need not make further mention of those points of interest here, as we shall have occasion to visit them all, under the head of one or other of the group of New England States, as we follow the net-work of routes by which they may be reached.

While the most thoughtless traveller will thus find, in the physical aspect of Massachusetts, ample sources of pleasure, the more earnest will not fail to draw yet higher delight from the strongly-marked *morale* of the country. Though small in area, compared with some other States of the confederacy, it is yet, in all the qualities which make national fame, one of the greatest of them all. Nowhere are there records of historical incident of higher sequence; nowhere a more advanced social position, or a greater intellectual attainment; nowhere a nobler spirit of commercial enterprise; nowhere a more inventive genius, a more indomitable industry.

In Massachusetts, more than in any other section of the Union, the dullest perception will be impressed with the evidences of all the highest and best characteristics of the American mind and heart, those wise and persistent qualities, which the Pilgrim Fathers first planted upon the shore at Plymouth, where the history of the State began, with the landing of the May Flower, on the memorable 22d of December, 1620 : the same righteous and unyielding nature which commenced the struggle for the national independence, in 1775, at Lexington and Bunker Hill.

RAILWAYS.—In a State so crowded with active and prosperous cities and homes as Massachusetts, and among a people of such wonderful will and energy and ambition, there is, of course, no lack of railway communication—that great modern test of national enterprise ; and so Massachusetts, while excelling all her sister States in every phase of industrial and mechanical achievement, has built (her area and population considered) more miles of railway than either of them. The iron tracks cover all the land, uniting all parts of the State to all others, and to every section of the Republic. We forbear to catalogue these routes at this point, as we shall follow them all, closely, in our visits to the several sections of the State.

ROUTES TO BOSTON FROM NEW YORK.

ROUTE 1. *Railway*—From the depot in Fourth av. cor. 27th st., via New Haven, Hartford, Springfield, and Worcester, 236 miles, twice a day; or by the new *Shore Line*, via New Haven, New London, Stonington, and Providence. A pleasant and very speedy route to the latter city.

ROUTE 2. *Stonington*—By steamer,

daily, from pier No. 2 North River, to Stonington, Ct.; thence by railway, via Providence, R. I.

ROUTE 3. *Fall River.*—Steamer, daily, at 5 P. M., from pier No. 3 North River, via Newport, R. I., to Fall River, Mass.; thence by railway.

ROUTE 4. *Norwich Line.*—Steamer, daily, from pier No. 18 (Cortlandt st.) North River, to Allyn's Point, Ct.; thence by railway, via Norwich and Worcester.

The most expeditious way between New York and Boston is that which we have marked No. 1, Railroad Routes—generally called the New Haven Line and the Shore Line. The time on these lines is between 8 and 9 hours, leaving one city in the morning and reaching the other in the afternoon, or leaving in the afternoon and arriving before midnight. All the other routes, by steamboat and railway, occupy the night, starting about 5 P. M., and arriving by dawn next day.

The New Haven route (No. 1), is upon the N. Y. and N. H. road for 76 miles, to New Haven all the distance along the south line of the State of Connecticut, near the shore of the Long Island Sound. To William's Bridge, 13 miles from New York, the track is the same as that of the Harlem R. R. to Albany. Leaving William's Bridge, we pass the pretty suburban villages of New Rochelle, Mamaroneck, Rye, and Port Chester, and reach Stamford 36 miles from New York. (The *Shore Line* leaves the other route at New Haven and extends through New London, Stonington, and Providence.)

Norwalk (New Haven Line—44 m.) is a pleasant village upon Norwalk river. The Norwalk and Danbury railroad, 24 miles, comes in at this point. The quiet, rural beauties of Norwalk make it one most available of the summer homes of Connecticut; particularly as it is scarcely beyond suburban reach of New York.

Bridgeport, 58 miles from New York, is the southern terminus of the Housatonic R. R., which traverses the valley of the Housatonic 110 miles to Pittsfield, Mass. This route is through the most picturesque portions of Connecticut and Massachusetts—the western or mountain regions. (See *Valley of the Housatonic.*) The Naugatuck R. R. extends hence via Waterbury to Winsted. Steamers ply between New York and Bridgeport. Bridgeport is upon an arm of the Long Island Sound, at the mouth of the Pequannock River. A terrace height of 50 feet, occupied by beautiful private mansions and cottages, commands a charming view of the town and the Sound. Iranistan, the picturesque residence of P. T. Barnum, formerly stood about 1½ mile west. It was entirely destroyed by fire.

New Haven, 76 miles from New York, is one of the most beautiful and most interesting places in New England. It is known as the City of Elms in a land of Elms, from the extraordinary number of beautiful trees of this species by which the streets are so gratefully shaded and so charmingly embellished. New Haven is a semi-capital of Connecticut. It is famous as the seat of Yale College, which has sent out more graduates than any other institution in America. The buildings of the College, which occupy nearly a square, are among the chief attractions of the city; especially the apartments devoted to the Fine Arts, and occupied by the large collection of the works of the eminent painter Trumbull. The *American Journal of Science and Art,* edited by Professors Silliman and Dana, and other literary periodicals, are published here. Steamboats connect the city with New York. The New Haven and Northampton or Canal R. R. extends 76 miles to Northampton, and the New Haven and New London R. R. 50 miles to New London. Population in Sept., 1853, was about 23,000.

Hartford.—Leaving New Haven our route turns northward from the Sound, over the New Haven, Hartford, and Springfield R. R. Hartford, a semi-capital of Connecticut, is 36 miles from New Haven and 112 from New York, and 124 from Boston. It is upon the right bank of the Connecticut River, navigable to this point by sloops and

State House and Yale College, (New Haven).

small steamboats, 50 miles up from Long Island Sound. Among the literary and educational institutions of Hartford are Trinity College, the Wadsworth Athenæum, the Connecticut Historical Society. Among its chief benevolent establishments, for which it is famous, are the American Asylum for the Deaf and Dumb, and the Retreat for the Insane. That old historic relic, the Charter Oak, held in so much reverence, stood in Hartford until 1856, when it was prostrated by a violent storm. Here is the residence of the distinguished poetess, Mrs. L. H. Sigourney. The population of Hartford in 1855, was about 20,000. Passengers from New York for Providence change cars here, and take the Providence, Hartford, and Fishkill R. R., 90 miles to Providence.

Springfield, Mass., is upon the Connecticut River, 26 miles north of Hartford, 98 miles from Boston, and 138 from New York. The U. S. Arsenal, located here, is the largest in the Union. It is charmingly perched upon Arsenal Hill, looking down upon the beautiful town, the river, and the fruitful valleys. This noble panorama is seen with still better effect from the cupola which crowns one of the arsenal buildings. This establishment employs nearly 300 hands, and 175,000 stands of arms are kept constantly on hand. This is a famous gathering point of railroads. The Connecticut Valley routes start hence, and furnish one of the pleasantest ways from New York to the White Mountains, through Northampton, Brattleboro', Bellows Falls, to Wells River and Littleton, N. H. (See "Valley of the Connecticut" and White Mountain routes, No. 10.) The Western railway from Albany to Boston passes through Springfield also, and continues our present route to Worcester. Population in 1850 about 20,000.

Worcester is a flourishing city of

The Stonington and the Fall River Routes.

20,000 people, 45 miles from Boston, in the centre of one of the most productive agricultural regions of Massachusetts. It is also a place of extensive manufactures. Quite a net-work of railways connects the city with all parts of the country. The Western road, direct from Boston to Albany; the Worcester and Nashua, communicating through other routes with the St. Lawrence River; the Worcester and Providence; the Norwich and Worcester, and the Boston and Worcester, which we now follow to the end of our present journey.

The Stonington Route, (No. 2.)—This route, as well as Nos. 3 and 4, by Fall River and by Norwich, takes us from New York by steamboat around the Battery and Castle Garden, along the whole eastern line of the city, and by the cities of Brooklyn and Williamsburg, up the beautiful East River by the suburban villages on the Long Island shore, by Blackwell's, Ward's, and Randall's Islands (covered by the public asylums and prisons), through the famous passage of Hell Gate, and up Long Island Sound—a gallery of admirable pictures, seen as they are from the Boston boats, in the declining evening light.

Stonington is in Connecticut, upon the coast near the eastern entrance to the Sound. Here we leave the steamer and take the railroad, 50 miles, through R. I. to Providence, thence to Boston, 43 miles, by the Boston and Providence R.R. (See Providence and vicinity.)

The Fall River Route, (No. 3.)—By steamer on Long Island Sound, round Point Judith, and up Narragansett Bay to Newport, R. I. (see Newport), and thence to Fall River, Mass. From Fall River 51 miles to Boston, by Old Colony and Fall River R. R.

Fall River is a thriving town of nearly 12,000 inhabitants and very extensive manufactures. It is at the entrance of Taunton River into Mount Hope Bay, an arm of the Narragansett. The historic eminence of Mount Hope,

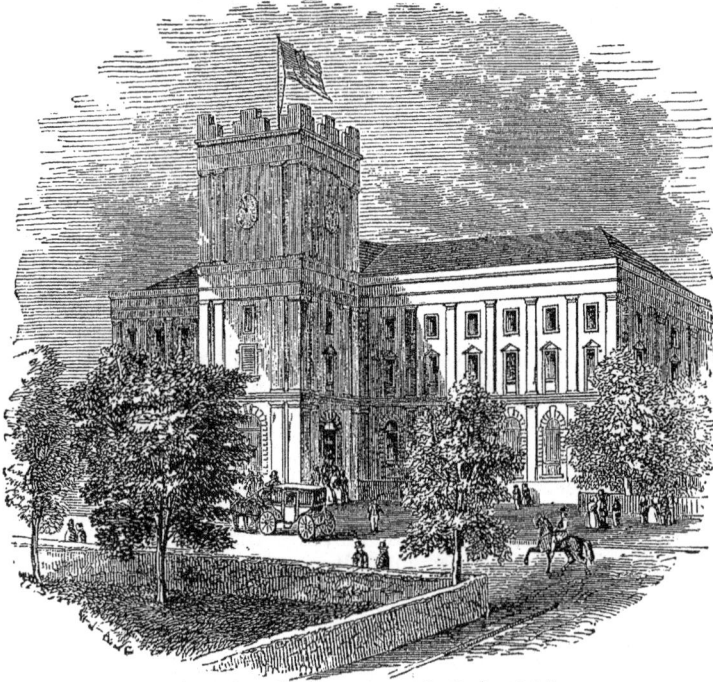

The United States Arsenal, Springfield.

SOUTH BOSTON

SOUTH BOSTON

PUBLIC GARDEN

COMMON
75 Acres

Liberty Tree

State House

Beacon

Chesnut

Byron

Charles

West Boston Bridge

Western Av.

Mill Dam or

PROVIDENCE RAIL ROAD

WORCESTER RAIL ROAD

Tremont

Marion

Knox

Boylston

Eliot

Carver

Washington

Bedford

Essex

Kneeland

Purchase

Fort Hill

BOSTON BAY

Brattle

Cornhill

State

School

Court

Water

Commercial

Faneuil Hall Market

Grave Yd.

First

Second

Broadway

Foundries

Iron St.

OLD COLONY R.R.

Turnpike

Granite

Boston

1 L
2 R
3 B
4 M
5 H
6 H
7 P
8 A
9
10
11
12
13
14
15
16
17
18
19

the home of King Philip, is admirably seen across the bay. Steamboats connect Newport with Providence by this route, via Narragansett Bay. *The Norwich Route*, (No. 4.)—This line is also by steamboat from New York, via Long Island Sound to the mouth of the Thames River, which it ascends to Allyn's Point, and passengers there take cars and follow the course of the Thames through Connecticut, directly north to Worcester; thence with other lines to Boston.

BOSTON AND VICINITY.

Boston is one of the most interesting of the great American cities, not only from its position as second in commercial rank to New York alone, but from its thrilling traditionary and historical associations, from the earliest days of discovery and colonization on the western continent; and through all the trials and triumphs of the childhood, youth, and manhood of the Republic—from its dauntless public enterprise, and from its high social culture and morals; from its great educational and literary facilities; from its numerous and admirable benevolent establishments; from its elegant public and private architecture, and from the surpassing natural beauty of all its suburban landscape.

Boston is divided into three sections, of Old Boston, East, and South Boston. The old city is built upon a peninsula of some 700 acres, very uneven in surface, and rising at three different points into eminences, one of which is 138 feet above the sea. The Indian name of this peninsula was Shawmut, meaning "Living Fountains." It was called by the earlier inhabitants Tremont or Trimountain, its sobriquet at the present day. The name of Boston was bestowed on it in honor of the Rev. John Cotton, who came hither from Boston in England. The first white inhabitant of this peninsula, now covered by Boston Proper, was the Rev. John Blackstone. Here he lived all alone until John Winthrop—afterwards the first Governor of Massachusetts—came across the river from Charlestown,

where he had dwelt with some fellow-emigrants for a short time. About 1635 Mr. Blackstone sold his claim to the now populous peninsula for £30, and removed to Rhode Island. The first church was built in 1632; the first wharf in 1673. Four years later a postmaster was appointed, and in 1704 (April 17), the first newspaper, called the "Boston News-Letter," was published.

A narrow isthmus, which is now called the Neck, joins the peninsula of Old Boston to the main land on the south, where is now the suburb of Roxbury, connected at this day with the city by numerous broad avenues. Many bridges link Charlestown, South Boston, and the main with the peninsula. These structures are among the peculiarities of the place, in their fashion, their number, and their length. The first one which was built was that over Charles River to Charlestown, 1,503 feet long. The Old Cambridge Bridge, across Charles River to Cambridge Road, is 2,758 feet in length, with a causeway of 3,432 feet. The South Boston Bridge, which leads from the Neck to South Boston, is 1,550 feet long. The Canal Bridge between Boston and Lechmere Point, is 2,796 feet, and from its centre another bridge extends 1,820 feet, to Prison Point, Charlestown. Boston Free Bridge to South Boston is 500 feet; and Warren Bridge to Charlestown is 1,390 feet. Besides these bridges, a causeway of a mile and a half extends from the foot of Beacon street to Sewell's Point, in Brookline. This causeway is built across the bay upon a substantial dam. Other roads lead into Boston over especial bridges, connecting the city with the main as closely as if it were a part thereof. Thus the topography of Boston is quite anomalous as a *mountain* city in the sea!

South Boston extends some two miles along the south side of the harbor, from Old Boston to Fort Independence. Near the centre, and two miles from the State House, are *Dorchester Heights*, the memorable battle ground where, in the Revolution, the enemy were driven from Boston. A fine view of the city, of the vicinity, and the sea, may be obtained from these Heights. Here, too, is a large reservoir of the Boston water works.

East Boston is upon the western part of Noddle's Island. This island was the homestead of Samuel Maverick, while John Blackstone was sole monarch of the peninsula, 1630. Here is the wharf, 1,000 feet long, of the Cunard line of British steamships. East Boston is the terminus, also, of the Grand Junction Railway. Chelsea is near by.

The streets of Boston, which grew up according to circumstances, are many of them very intricate, and troublesome to unravel, a difficulty which is being gradually obviated in a degree. The fashionable promenades and shopping avenues are, first, Washington street and then Tremont street.

Boston Common is a large and charming public park in the old city, and is, very justly, the pride of the people and the admiration of strangers. It contains nearly 50 acres, of every variety of surface, up-hill and down, and around, all covered with inviting walks, grassy lawns, and grand old trees. A delicious pond and fountain occupy a central point in the grounds, and around them are many of the old mansions of the place—led, on the upper

Boston Common.

hill, by the massive, dome-surmounted walls of the State Capitol. The Common drops from Beacon street, the southern declivity of Beacon Hill, by a rapid descent to Charles River.

Faneuil Hall. This famous edifice, called the " Cradle of Liberty," is in Dock square. It is about 109 years old, and is an object of deep interest to Americans. Here the fathers of the Revolution met to harangue the people on the events of that stirring period ; and often since that time the great men of the State and nation have made its walls resound with their eloquence. It was presented to the city

Faneuil Hall.

by Peter Faneuil, a distinguished merchant, who, on the 4th of July, 1740, made an offer, in a town-meeting, to build a market-house. There being, at that time, none in the town, it was, as a matter of course, accepted. The building was begun the following year, and finished in 1742. The donor so far exceeded his promise, as to erect a spacious and beautiful Town Hall over it, and several other convenient rooms. In commemoration of his generosity, the town, by a special vote, conferred his name upon the Hall; and, as a further testimony of respect, it was voted that Mr. Faneuil's full-length portrait be drawn at the expense of the town, and placed in the Hall. This, with other pictures, can be seen by visitors.

The State House is on the summit of Beacon Hill, and fronting the "Com-mon." Its foundation is 110 feet above the level of the sea. Length 173 feet, breadth 61. The edifice was completed in 1798, at a cost of $133,330, about three years having been occupied in its construction. On the entrance floor is to be seen Chantrey's statue of Washington. Near by is the staircase leading to the dome, where visitors are required to register their names, and from the top of which is obtained a fine view of the city, the bay, with its islands, and the country around.

The Exchange, on State street, was completed in the fall of 1842. It is 70 feet high and 250 feet deep, covering about 13,000 feet of ground. The front is built of Quincy granite, with four pilasters, each 45 feet high, and weighing 55 tons each. The roof is of wrought iron, and covered with galvanized sheet iron; and all the principal staircases are fire-proof, being constructed of stone and iron. The centre of the basement story is occupied by the Post Office. The great central hall, a magnificent room, is 58 by 80 feet, having 18 very beautiful columns in imitation of Sienna marble, with Corinthian capitals, and a sky-light of colored glass, finished in the most ornamental manner. This room is used for the merchants' exchange and subscribers' reading room.

The Custom House is located at the foot of State street, between the heads of Long and Central wharves. It is in the form of a cross; the extreme length being 140 feet, breadth 95 feet. The longest arms of the cross are 75 feet wide, and the shortest 67 feet, the opposite fronts and ends being alike. The entire height to the top of the dome is 90 feet.

The Court House, a fine building in Court square, fronting on Court street, is built of Quincy granite.

The City Hall is near the Court House, and fronting on School street, with an open yard in front. Here, in September, 1856, a colossal bronze statue of Benjamin Franklin, who was a native of Boston, was erected, with great public parade and rejoicing. This fine work was modelled by R. B. Greenough, Esq., a brother of the distinguished sculptor, Horatio Greenough.

The Massachusetts Hospital covers an area of four acres on Charles River, between Allen and Bridge sts. Near by, at the foot of Bridge street is the Mass. Medical College. The Boston Music Hall fronts on Winter street and Bumstead place.

The Boston Athenæum occupies an imposing edifice on Beacon near Tremont street. It was incorporated in 1807, and is one of the best endowed literary establishments in the world. There are in the library 50,000 volumes, and an extensive collection of manuscripts. The Athenæum possesses a fine gallery of paintings, in connection with which the annual displays of the Boston artists are made.

The Massachusetts Historical Society, organized in 1794, possesses 12,000 volumes, and many valuable manuscripts, coins, charts, maps, &c. The Boston Public Library is on Boylston street. It possesses, at this time, about 80,000 vols. The American Academy of Arts and Sciences, one of the oldest societies of the kind in the country, has 15,000 volumes. It occupies an apartment in the Athenæum. Besides these libraries Boston has many others, as, the State Library, the Bowditch, the Social Law Library, &c.

The Lowell Institute provides for regular courses of free lectures upon natural and revealed religion, and many scientific and art topics. We may mention, also, among literary, scientific, and art societies of the city, the Lyceum, the Natural History, the American Oriental, the American Statistical, the Musical Educational, and the Handel and Haydn Societies, and the Boston Academy of Music.

Harvard University. This venerable seat of learning is at Cambridge, three miles from the city of Boston. It was founded in 1638, by the Rev. John Harvard. The University embraces, besides its collegiate department, law, medical, and theological schools. The buildings are 15 in number, all located in Cambridge, except that of the medical school in North Grove street, in Boston.

The *Old Washington Head-Quarters*, at Cambridge, known as the Cragie House, where the poet Henry W. Longfellow has resided for many years, is near the Harvard University. It was at Cambridge that the painter, Washington Allston, lived and died.

Boston, always so much distinguished for its literary character, as to have won the name of the Athens of America, has, besides its innumerable libraries and institutions of learning, more than a hundred periodical publications, and newspapers, dealing with all themes of study, and all shades of opinion and inquiry.

The churches of the city are numerous, as might be expected of the home

Bunker Hill Monument.

of the Puritans. They are more than 100 in number—the Unitarians having the largest share. Many of the churches are very costly and imposing edifices. The oldest is Christ Church, built in 1723, and the next is the famous " Old South," erected in 1730. This is a building of great historical interest.

Hotels. The most fashionable are the Tremont House, on Tremont street; the Revere House, on Bowdoin square; the Winthrop House, Tremont street; the American House, Hanover street; the Adams; the United States, &c,

Theatres. The Boston Theatre, 361 Washington street; the Howard Athenæum, Howard street, the Museum, Tre-

mont street; the National Theatre, corner of Portland and Travers street; Ordway Hall, &c.

Bunker Hill Monument, commemorative of the eventful battle fought on the spot, is in Charlestown. The top of this structure commands a magnificent view, embracing a wide extent of land and water scenery. The journey up is somewhat tedious, traversing nearly 300 steps —yet this is forgotten in the charming scene and delightful air which repay the visitor. Near at hand is the United States Navy Yard, containing, among other things, a rope-walk—the longest in the country.

Mount Auburn Cemetery, about a

mile from Harvard University, and about four from Boston, by the road from Old Cambridge to Watertown. It is the most beautiful of American rural burying-places, embellished by landscape, and horticultural art and taste, and by a most picturesque chapel, and many elegant and costly monuments. Its walks, and lanes, and lawns make it the most delightful of all the resorts in the vicinage of the city. Cars run from the station in Bowdoin square, every 15 minutes, during the day, and until half-past eleven o'clock at night. *Mr. Cushing's Garden*, a place of great beauty, is a short distance beyond Mount Auburn, in Watertown. Tickets may be obtained gratis on application at the Horticultural Store in School street. —*Fresh Pond*, another charming place of resort, is about four miles from Boston, and half a mile from Mount Auburn; it is reached by the railroad cars, which leave Charlestown many times during the day. *Fare* 12½ cents.— The other fine sheets of water in the vicinity of Boston, well worthy the attention of visitors, are *Horn, Spot, Spy,* and *Mystic* Ponds.

WATERING AND OTHER PLACES IN THE VICINITY OF BOSTON.

Nahant, a delightful watering-place, is situated about 12 miles from Boston, by water, and 14 by land. During the summer season, a steamboat plies daily. *Fare* 25 cents. This is a most agreeable excursion, affording an opportunity, in passing through the harbor, for seeing some of the many beautiful islands with which it is studded. Nahant may also be reached by taking the Eastern Railway cars as far as Lynn, and thence walking or riding a distance of three miles, along the hard sandy beach, in full view of the open sea; or by omnibuses, which intersect the railroad cars, at Lynn, several times daily.

The peninsula is divided into Great and Little Nahant, and Bass Neck. The former is the largest division, containing 300 acres—a part of which is under cultivation—many handsome dwellings, and a spacious hotel, with a piazza on each floor. From this place the visitor has a boundless sea-coast view.

On the south side of Great Nahant is the dark cave or grotto, called the *Swallow's Cave*, 10 feet wide, 5 feet high, and 70 long, increasing in a short distance, to 14 feet in breadth, and 18 or 20 in height. On the north shore of the peninsula is a chasm 20 or 30 feet in depth, called the *Spouting Horn*, into which, at about half tide, the water rushes with great violence and noise, forcing a jet of water through an aperture in the rock to a considerable height in the air.

Philip's Beach, a short distance north-east of Nahant, is another beautiful beach, and a noted resort for persons in search of pleasure or health.

Nantasket Beach, 12 miles from Boston, is situated on the east side of the peninsula of Nantasket, which forms the south-east side of Boston Harbor, and comprises the town of Hull. The beach, which is remarkable for its great beauty, is four miles in length, and celebrated for its fine shell-fish, sea-fowl, and good bathing.

Chelsea Beach, about three miles in length, is situated in the town of Chelsea, and is another fine place of resort. A ride along this beach on a warm day is delightful. It is about five miles from Boston, and may be reached by crossing the ferry to East Boston.

Lynn is nine miles from Boston, on the Eastern Railroad. It is charmingly situated on the north-east shore of Massachusetts Bay, in the immediate vicinity of Nahant, and is a famous place for the manufacture of ladies' shoes. This business, here, employs 150 establishments and 10,000 hands, half of whom are females. It is estimated that 4,500,000 pairs of ladies' and misses' shoes are made here every year, amounting, in value, to $3,500,000. Besides the product of the city, another half a million pairs are made in the neighborhood.

Salem is a beautiful city, 16 miles from Boston, by the Eastern Railroad. It extends about two miles along and three quarters of a mile across the peninsula formed by the north and the

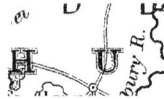

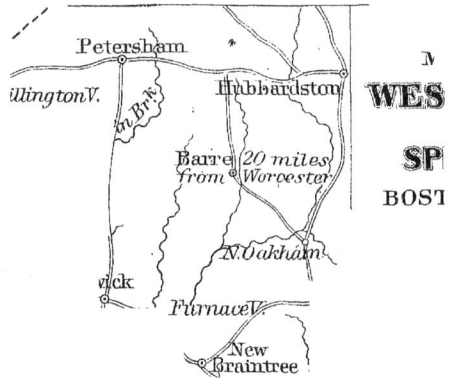

CONTINUATION OF THE
WESTERN R.R. TO ALBANY,
AND THE
HOUSATONIC TO HUDSON.
ALSO THE
CONNECTICUT RIVER
ROUTE &c.

Wil
to

nts
ill

Gree

Sth

Williamstown

Adan
13

New
Ashford

Highland

PITTSF. & N.ADA

Che
9

E L A E R

Alps V.   N. Stephentown

Hancock

Lanesborough

5

E W

assau

E. Nassau V.

Stephentown

nesborough

R   E

New Lebanon

Springs

Wes

N. Britain

Shaker V.

Shaker
V.
156

67 Canaan

Kinder-
hook

Stuyve-
sant

Humph-
reys F

Grade at Canaa

Chatham
Corners 177

162

St

Y O R K

BERKS. R.R

Chittenden

Stephen-
town

W. Stockbridge
Centre

olumbia
ville

Ghent 181

Green River V.

ee

Sth

Gle

Green
water P.

B

Waterville

ument

C O L U M B I A

HUDSON

ns
HUDSON
9

Cloverack

Hillsdale

Alfo

Hop Brk

nville

K

Hop Bk V.

Claverack

Copake

Smoky Hollow

York
V.

Tyringham

Otis

Glenco

R

Johnstown

Taghkanic

Copake
P.

Harts V.

Millers V.

Clermont

Charlotte

Copake

Ancram

Mt
Washington

Boston Corne

Sandisfield

field

New

Shooks V.

Places in the vicinity of Boston.

south rivers. It is distinguished for its literary institutions, and for its commercial enterprise. Next to Plymouth, it is the oldest town in New England. Salem was the chief scene of the "witchcraft" madness in 1692. Upon Gallows Hill—a fine eminence overlooking the city—19 persons of the town and the neighborhood were executed for this supposed crime.

Salem is also distinguished for its services in the war of the Revolution. *Marblehead* is 18 miles from Salem, by a branch road.

Beverly is upon an arm of Ann Harbor, two miles from Salem, with which it is connected by a bridge of 1,500 feet, and from Boston 18 miles, via Eastern Railroad.

Wenham, 22 miles from Boston, via Eastern Railway. Wenham Pond, a beautiful sheet of water, about a mile square, affords abundance of excellent fish, and is much visited by persons fond of angling. It is also noted for the quality and quantity of its ice, a large amount of which is yearly exported.

Newburyport, Mass., 36 miles from Boston, via Eastern Railroad, lies on a gentle acclivity, on the south bank of the Merrimac River, near its union with the Atlantic. It is considered one of the most beautiful towns in New England. In consequence of a sand-bar at the mouth of the harbor, its foreign commerce has greatly declined. The celebrated George Whitefield died in this town in September, 1770.

Salisbury Beach, celebrated for its beauty and salubrity, is much visited during the warm season. It is from four to five miles distant from Newburyport.

Hampton, 43 miles from Boston, via Eastern Railroad, is pleasantly situated near the Atlantic coast. From elevations in the vicinity there are fine views of the ocean, the Isle of Shoals, and of the sea-coast from Cape Ann to Portsmouth. Hampton Beach has become a favorite place of resort for parties of pleasure, invalids, and those seeking an invigorating air. Great Boar's Head, in this town, is an abrupt eminence extending into the sea, and dividing the beaches on either side. There is here an hotel for the accommodation of visitors. The fishing a short distance from the shore is very good.

The Isle of Shoals is distant about nine miles from Hampton and from Portsmouth. These shoals are seven in number. Hog Island, the largest, contains 350 acres, mostly rocky and barren. Its greatest elevation is 59 feet above high-water mark. Upon this island is a hotel, recently erected. These rocky isles are a pleasant resort for water parties, and the bracing air, while refreshing to the sedentary, cannot be otherwise than salutary to invalids.—*Rye Beach* is another noted watering-place on this coast, much frequented by persons from the neighboring towns.

Portsmouth, N. H., 56 miles from Boston, and 51 from Portland, Maine, by the Eastern Railroad, the principal town of the State, and the only seaport, is built on the south side of the Piscataqua River. Its situation is a fine one, being on a peninsula near the mouth of the river. It is connected by bridges with Kittery in Maine, and Newcastle on Grand Island, at the mouth of the river. The harbor is safe and deep, and is never frozen, its strong tides preventing the formation of ice. There is here a United States Navy Yard, one of the safest and most convenient on the coast. The North America, the first line-of-battle-ship launched in this hemisphere, was built here during the Revolution.

Andover, the seat of Phillips' Academy, and of the Andover Theological Seminary, is 23 miles from Boston, on the Boston and Maine Railway.

Lowell. This famous manufacturing city, the first in the Union, is upon the Merrimack, 26 miles from Boston, by the Boston and Lowell Railroad. Lowell was incorporated as a town in 1826, and in 1854 its population was about 37,000. There are over 50 mills in operation in Lowell, employing a capital of $13,900,000, and nearly

13,000 hands, of whom about 9,000 are females.

Concord is situated on the river of the same name, 20 miles from Boston, by the Fitchburg Railroad. It is celebrated as the place where the first effectual resistance was made, and the first British blood shed, in the Revolutionary war. On the 19th of April, 1775, a party of British troops was ordered by General Gage to proceed to this place to destroy some military stores, which had been deposited here by the province. The troops were met at the north bridge by the people of Concord and the neighboring towns, and forciby repulsed. A handsome granite monument, erected in 1836, commemorates the heroic and patriotic achievement.

Lexington, the scene of the memorable battle of Lexington, at the commencement of the Revolution, April 19, 1775. From Boston bv Fitchburg and branch railways through West Cambridge—a fine ride of eleven miles; fare, 35 cents.

Brighton is five miles from Boston, on the Boston and Worcester Railroad. This is a beautiful suburban town, on the south side of Charles River. It is also a noted cattle market.

Quincy is eight miles from Boston, by the Old Colony and Fall River Road. Famous for its granite quarries; and as the birth-place of John Hancock, Josiah Quincy, Jr.; Presidents John and John Quincy Adams, and other eminent men. The fine estate of the Quincy family is here.

Plymouth is 37 miles from Boston, by the Old Colony and Fall River and branch roads. It is a spot of especial interest, as the landing-place of the Pilgrim Fathers, and as the oldest town in New England. The immortal Plymouth Rock lies at the head of Hedge's Wharf. It is now much reduced from its ancient proportions, being only 6½ feet across its greatest breadth, and but 4 feet thick. The surface only is visible above the ground. The landing of the Pilgrims from the May Flower, occurred on the 22d December, 1620.

Marshfield, interesting as the home of Webster, is 28 miles south-east of Boston, by the Old Colony and the South Shore Railways.

Hingham is 17 miles from Boston, by the Old Colony and the South Shore Railways; or may be reached by a pleasant sail down the bay.

· **Cohassett,** three miles from Hingham (South Shore Road), is a popular sea-side resort.

Taunton, Mass., is a beautiful town of some 12,000 inhabitants, situated at the head of navigation, on the Taunton River. It may be reached from Boston, 35 miles; and from Providence, 30 miles, by the Boston and Providence Railroad, taking the New Bedford and Taunton Road at Mansfield, about midway between Boston and Providence.

New Bedford is a charming maritime city, of some 18,000 people; situated on an estuary of Buzzard's Bay. It is famous for its whale fisheries, in which enterprise it employs between 300 and 400 ships. New Bedford is the terminus of the New Bedford and Taunton Railway, by which route, via Mansfield, on the Boston and Providence Road, it may be reached from those two cities. It is accessible also from New York and Boston, by the Fall River route. Distance from Boston, 55 miles.

Martha's Vineyard, and Nantucket. These famous sea-islands lie off New Bedford, with which port they are in daily steamboat communication. Nantucket may be still more easily reached via Cape Cod Railway to Hyannis.

Middleborough is a prosperous town on the Fall River route, between New York and Boston, at its point of junction with the Cape Cod Railroad. It is pleasantly situated upon the Taunton River, 40 miles from Boston. It is the seat of a very popular Scholastic Institution, under the direction of the Rev. Mr. Jenks.

Cape Cod, and the Sea Islands. Those who delight in the sea breezes, in salt water bathing, and fishing, and in the physical beauties and wonders of the ocean changes, will find ample gratification everywhere upon the Atlantic borders, and especially upon the bold islanded coast of New England.

Besides the well-known haunts of the Long Island and the Jersey shores, of Newport, and of the numerous suburban resorts of Boston, to which we have elsewhere alluded, the Isle of Shoals, off Portsmouth, Martha's Vineyard, and Nantucket, off New Bedford, &c., we commend the summer wanderer to a tour through the towns and villages, and along the coasts of that very secluded portion of Massachusetts—Cape Cod. Let him journey from "Plymouth Rock," the inner point, to Province Town, the outer verge, and he will find novelties in both physical nature and social life, which will be most agreeable. The Cape Cod Railway from Boston will take him far along upon the Cape, 74 miles to Barnstable ; 76 to Yarmouth ; and 80 to Hyannis ; or he may go thence by steamboat, and afterwards continue from point to point by stage.

THE CONNECTICUT VALLEY, RIVER, AND RAILWAYS.

The beautiful valleys watered by the Connecticut, are among the most inviting portions of the New England landscapes ; whether for rapid transit, or for protracted stay. The whole region is speedily and pleasantly accessible from every point, and may be traversed *en route*, to most of the principal summer resorts of New England, since many important and very attractive towns and villages lie within its area, and since it is crossed, and recrossed, every where, by the intricate railway system, which unites Boston so intimately, not only with all the Eastern States, but by connection infinite, with the whole country.

The Connecticut, the Queen of New England rivers—the chiefest and most beautiful—rises in the hills of New Hampshire and Vermont, near the Canada borders ; and flowing nearly southward, for 400 miles, separates the two States of its mountain birth ; traverses the entire breadth of Massachusetts and Connecticut, to the Long Island Sound.

Its waters are swelled by the tribute of the Passumpsic, the White, the Deerfield, the Westfield, the Ammanoosuc, and other rivers. It is navigable for sloops 50 miles up to Hartford, and with the help of numerous canals, very much farther. The Connecticut Valley is, perhaps, 300 miles long in a straight line, with a mean width of 40 miles. The soil is as fertile as the landscape is beautiful.

Railroads from New York, Boston, Albany, and other places, meet at Springfield, the southern threshold of the most picturesque part of the Connecticut ; so we will commence our tour here—referring the traveller to the route from New York to Boston, for mention of Springfield itself.

Northampton, Mass., is 17 miles above Springfield, on the line of railway which follows the Connecticut, up to the diverging lines for the White Hills of New Hampshire, and for Vermont and Canada. It is in every way one of the most charming villages in New England, and none other is more sought for summer residences. It lies about a mile west of the Connecticut, surrounded by rich alluvial meadows, sweeping out in broad expanse, from the base of grand mountain ridges. The village is not too large for country pleasures, the population of the township falling within 6,000 ; yet, its natural advantages are so great, and so many pleasant people have established themselves here in such pleasant and beautiful places, and the hotels are so admirable, that the tourist, the most *difficile*, will not miss either the social or the physical enjoyments of his city home. Even the little business part of Northampton has a cosy, rural air, and all around are charming villas, nestled on green lawns, and among fragrant flowers. Among the specialities of Northampton, are several distinguished water-cure establishments, the chief of which is that known as Round Hill, a large and beautiful place, upon the fine eminence after which it is named, just west of the village. The schools here have alway been in very high repute. Its chief academy is the Collegiate In-

4

stitute. The vicinage of Northampton is, perhaps, the most beautiful portion of the Connecticut Valley, the most fertile in its interval lands, and the most striking in its mountain scenes; for it looks out, directly, upon the crags and crests of those famous hills—Mount Holyoke and Mount Tom.

The hotels are, first, the Mansion House, an elegant establishment, upon the upper edge of the village; then Warner's Hotel, in the business street; and the Nonotuck House at the railway station. Northampton is united to New Haven, by the New Haven and Northampton Railroad, 76 miles long, as well as via Springfield.

Mount Holyoke is directly across the river from Northampton; a good carriage road winds to the summit, 1,120 feet above the sea, where there is a little inn and an observatory. There are not of its kind many scenes in the world more beautiful than that which the visitor to Mount Holyoke looks down upon; the varied features of the picture—fruitful valleys, smiling villages and farms, winding waters; and, afar off, on every side, blue mountain peaks, innumerable, will hold him long in happy contemplation.

"Mount Holyoke," says Mr. Eden's Handbook to the region, "is a part of a ridge of greenstone, commencing with West Rock near New Haven, and proceeding northerly across the whole of Connecticut; but its elevation is small until it reaches Easthampton, when it suddenly mounts up to the height of nearly 1,000 feet, and forms Mount Tom. The ridge crosses the Connecticut, in a northeast direction, and curving still more to the east, terminates 10 miles from the river, in the north-west part of Belchertown. All that part of the ridge east of the river is called Holyoke, though the Prospect House is erected near its south-western extremity, opposite Northampton, and near the Connecticut. This is by far the most commanding spot on the mountain, though several disitnct summits, that have as yet received no uniform name, afford delightful prospects.

"Nothing can be seen, except an occasional glimpse through the trees, in the ascent, until the visitor arrives at the Mountain House, and here the sudden burst of such a magnificent prospect is as startling as it is delightful. The visitor finds himself lifted up nearly 1,000 feet from the midst of a plain, which, northerly and southerly, is of great extent; and so comparatively narrow is the naked rock on which he stands, that he wonders it has withstood the winds and storms of so many centuries.

"Of all the charming objects, however, with which the landscape abounds, the most enchanting is the Connecticut itself. This stream may, perhaps, with as much propriety as any in the world, be named the Beautiful River. Joel Barlow, in his poem of the Columbiad, speaking of the Connecticut, says:—

' No watery glades thro' richer valleys shine,
    Nor drinks the sea a lovelier wave than
    thine.'

"Indeed, during its whole course, it uniformly sustains this character. The purity, salubrity, and sweetness of its waters, the frequency and elegance of its meanders, its absolute freedom from all aquatic vegetables; the uncommon and universal beauty of its banks, here, a smooth and winding beach; now fringed with bushes; now crowned with lofty trees, and now formed by the intruding hill, the rude bluff, and the shaggy mountain; are objects which cannot be thoroughly described or adequately imagined. The river turns four times to the east, and three times to the west, within 12 miles, and within that distance, makes a progress of 24. It winds its way majestically, yet most beautifully, through the meadows of Hatfield, Hadley, and Northampton, and directly in front of Holyoke, it formerly swept around in a graceful curve of three miles, without advancing in its oceanward course 100 rods; but in the spring of 1840 (as if impelled by the go-ahead character of the age), it cut across the neck of this peninsula, though as it still continues to pass around the curve, as well as through the new channel, the beauty of the spot is unimpaired. After this, it passes directly through the deep opening between Holyoke and Tom, which its own waters or some other agencies have excavated in early times. Below this point, the Connecticut is in full view, like a serpentine mirror, for nearly 20 miles.

"The intervals which in this view border it in continual succession, are fields containing from 500 to 5,000 acres, formed like terraced gardens, lowest near the river, and rising as they recede from it by regular gradations. These fields are distributed into an immense multitude of lots, geometrically diversified in the summer with grass, corn, grain, and other products of laborious industry. On the west, and a little elevated above the general level, the eye turns with delight to the populous village of Northampton; exhibiting in its public edifices and private dwellings, an unusual degree of neatness and elegance. A little more to the right, the quiet and substantial villages of Hadley and Hatfield, and still farther east and more distant, Amherst, with its College, Observatory, Cabinet, and Academy, on a commanding eminence, form pleasant resting places for the eye. The valley on the south of Holyoke is not as interesting as that on the west and north, chiefly because the land is less fertile. The village of South Hadley with Mount Holyoke Female Seminary, is indeed a pleasing object, but Springfield, though finely situated, is too far removed for an exhibition of its particular features. Other places south of Springfield are indistinctly visible along the banks of the Connecticut: and even the spires of some of the churches in Hartford may be seen in good weather, just rising above the trees. Still farther south may be seen the abrupt greenstone bluffs midway between Hartford and New Haven; and looking with a telescope between these, other low hills may be indistinctly seen, which may be the trap ridge encircling New Haven.

"Facing the south-west, the observer has before him on the opposite side of the river, the ridge called Mount Tom, rising 100 or 200 feet higher than Holyoke, and dividing the Valley of the Connecticut longitudinally. The western branch of this valley is bounded on the west by the Hoosac range of mountains; which, as seen from Holyoke, rises ridge above ridge for more than twenty miles, chekered with cultivated fields and forests, and not unfrequently enlivened by villages and church spires. In the north-west, Graylock may be seen peering above the Hoosac; and still farther north, several of the Green Mountains in Vermont, shoot up beyond the region of the clouds in imposing grandeur. A little to the south of west, the beautiful outline of Mount Everett is often visible. Nearer at hand, and in the valley of the Connecticut, the insulated Sugar Loaf and Mount Toby present their fantastic outlines; while, far in the north-east, ascends in dim and misty grandeur the cloud-capt Monadnoc.

"Probably, under favorable circumstances, not less than thirty churches, in as many towns, are visible from Holyoke. The north and south diameter of the field of vision there cannot be less than 150 miles.

"Mount Holyoke commands also a view of the Connecticut River Railroad for many miles; and it is a novel and interesting sight to witness the Iron Horse suddenly emerging from the Willimantic Bridge, and pursuing its impetuous and resistless course along the picturesque bank of the stream, and through the beautiful meadows below us."

Mount Tom, upon the opposite side of the river, is not yet so much visited as are its neighboring cliffs of Holyoke, though it is considerably higher, and the panorama from its crest is no less broad and beautiful.

The village of Easthampton (again quoting Mr. Eden's Guide), is situated on the west side of Mount Tom, four miles from Northampton. It contains a very extensive button manufactory, well deserving of a visit from those who

Mount Tom, Mass.

can appreciate mechanical ingenuity. The principal feature of the place, however, is its noble seminary for the youth of both sexes, which was founded and liberally endowed by the Hon. Samuel Williston, at an expense of $55,000, and has been in successful operation 15 years, having now an average attendance of about 200 pupils.

On the east side of Mount Tom and of the river is the village of South Hadley, famous as the seat of the Mount Holyoke Female Seminary, founded and for many years conducted by Miss Mary Lyon. This institution has sent out hundreds of graduates, as teachers, into all parts of the land. South Hadley has many spots which afford most agreeable prospects. Standing on the elevated bank of the river and facing the north-west, you look directly up the Connecticut, where it passes between Holyoke and Tom; those mountains rising with precipitous boldness, on either side of the valley through the opening, the river is seen for two or three miles, enlivened by one or two lovely islands, while over the rich meadows that constitute the banks, are scattered trees, through which, half hidden, appears in the distance the village of Northampton, its more conspicuous edifices being only visible.

The village of Hadley is connected with Northampton by a bridge over the Connecticut. The river immediately above the town, leaving its general course, turns north-west: then, after winding to the south again, turns directly east; and thus, having wandered five miles, encloses, except on the east, a beautiful interval, containing between two and three thousand acres. On the isthmus of this peninsula lies the principal street—the handsomest, by nature, in New England. It is a mile in length, running directly north and south; is sixteen rods in breadth; is nearly a perfect level; is covered during the fine season with a rich verdure; abuts at both ends on the river, and yields every where a delightful prospect.

There is a Grammar School here, that owes its foundation to funds left by the Hon. Edward Hopkins, a former Governor of Connecticut.

In this town resided for fifteen or sixteen years Whalley and Goff, two of those who composed the court for the trial of King Charles the First, and who signed the warrant for his execution. They came to Hadley in 1664. When the house which they occupied was pulled down, the bones of Whalley were found buried just without the cellar wall, in a kind of tomb formed of mason work, and covered with flags of hewn stone. After Whalley's death, Goff left Hadley and went, it was thought, to New York, and finally to Rhode Island, where he spent the rest of his life with a son of his deceased *confrère*.

Amherst, the seat of the famous College, is built upon an eminence, four miles east of Hadley. The College Observatory and especially its rich cabinet, should receive due consideration from the visitor here.

The Sugar-Loaf Mountain comes now into view, as we journey on up the valley. This conical peak of red sandstone rises almost perpendicularly five hundred feet above the plain, on the bank of the Connecticut, in the south part of Deerfield. As the traveller approaches this hill from the south, it seems as if its summit were inaccessible. But it can be attained without difficulty on foot, and affords a delightful view on almost every side. The Connecticut and the peaceful village of Sunderland on its bank, appears so near, that one imagines he might almost reach them by a single leap. This mountain overlooks a spot which was the scene of the most sanguinary conflicts that occurred during the early settlement of this region. A little south of the mountain the Indians were defeated in 1675 by Captains Lathrop and Beers; and one mile north-west, where the village of Bloody Brook now stands (which derived its name from the circumstance), in the same year, Captain Lathrop was drawn into an ambuscade, with a company of "eighty young men, the very flower of Essex County," who were nearly all destroyed.

The spot where Captain Lathrop and about thirty of his men were interred, is marked by a stone slab; and a marble monument, about twenty feet high and six feet square, is erected near by.

Deerfield Mountain rises some 700 feet above the plain on which the village stands. From the western verge of this summit the view is exceedingly interesting.

The alluvial plain on which Deerfield stands is sunk nearly 100 feet below the general level of the Connecticut valley; and at the south-west part of this basin, Deerfield river is seen emerging from the mountains, and winding in the most graceful curves along its whole western border. Still more beneath the eye is the village, remarkable for regularity, and for the number and size of the trees along the principal street. The meadows, a little beyond, are one of the most verdant and fertile spots in New England. Upon the whole, this view is one of the most perfect rural peace and happiness that can be imagined.

A few miles north of Deerfield, and in the same valley, but on higher ground, can be seen the lovely village of Greenfield. As we approach this place from the south, the view is one of great beauty.

"How gay the habitations that bedeck
This fertile valley! Not a house but seems
To give assurance of content within;
Embosomed happiness and placid love;
As if the sunshine of the day were met
By answering brightness in the hearts of all
Who walk this favored ground."

Mount Toby lies in the north part of Sunderland, and west part of Leverett, and is separated from Sugar-Loaf and Deerfield Mountains by the Connecticut river. On various parts of the mountain, interesting views may be obtained, but at the southern extremity of the highest ridge, there is a finer view of the valley of the Connecticut than from any other eminence. Elevated above the river nearly 1,000 feet, and but a little distance from it, its windings lie directly before you; and the villages that line its banks—Sunderland, Hadley, Hatfield, Northampton, and Amherst, appear like so many sparkling gems in its crown.

Mount Warner is a hill of less altitude than any before named, being only 200 or 300 feet in height, but a rich view can be had from its top of that portion of the valley of the Connecticut just described. It lies in the north part of Hadley, not more than half a mile from the river, and it can be easily reached by a carriage. A visit to it can, therefore, be performed by the invalid, and will form no mean substitute for an excursion to Holyoke or Toby.

Greenfield, in its business quarter, is a lively little place. The wonted New England quiet, however, is all around it, in elm-shaded streets and garden-surrounded villas. The high hills in the neighborhood open fine pictures of the valleys and windings of the great river.

Greenfield is the terminus of a railway from Boston, via Fitchburg, 100 miles from the former, and 56 miles from the latter place   Other routes now unite it with the railway systems of the west and of the north-west. The Green River, which flows near the village, is a pretty stream, and hard by are the Deerfield and Greenfield rivers. Among the manufactures of Greenfield there is a tool shop, in which are made 382 different shapes of carpenter's planes. In an extensive cutlery establishment upon Green River, 300 operatives are employed.

Vernon.—At Middle Vernon there is a charming view up the river, as seen from the railway track · Mount Chesterfield, in New Hampshire, opposite Brattleborough, rising up stoutly in the back-ground.

Brattleborough brings us fairly out of the rich alluvial lands into the upper and more rugged portions of the Connecticut. The intervals now grow narrower, and the hills more stern. This beautiful village is in a very picturesque district, upon the west side of the river. It is, deservedly, one of the most esteemed of the summer resorts of the Connecticut, so pure and health-restoring are its airs, and so pleasant all

Mount Chesterfield, N. H.

its belongings, within and without. There are here several large and admirable water-cure establishments, and a fine first-class hotel, called the Revere House. The village cemetery on a lofty terrace overlooking the river, above and below, is a beautiful rural spot. West River, above the town, is an exceedingly picturesque stream. The buildings and grounds, in this vicinity, of the Asylum for the Insane, have a fine manorial appearance.

Our next stage is 24 miles, from Brattleborough to Bellows Falls, over the Vermont valley road.

Bellows' Falls is a famous congregating and stopping place of railways. With the exception of some bold passages of natural scenery, and a most sumptuous summer hotel, called the Island House, there is not much here, comparatively, to allure the traveller. Railways come in from Boston on the east, from the valley of the Connecticut on the south, from Vermont and Canada on the north, and from Albany and Troy, via Rutland, on the west.

The Falls are a series of rapids in the Connecticut, extending about a mile along the base of a high and precipitous hill, known as " Fall Mountain," which skirts the river on the New Hampshire side. At the bridge which crosses the river at this place, the visitor can stand directly over the boiling flood; viewed from whence, the whole scene is effective in the extreme. The Connecticut is here compressed into so narrow a compass that it seems as if one could almost leap across it. The water, which is one dense mass of foam, rushes through the chasm with such velocity, that in striking on the rocks below, it is forced back upon itself for a considerable distance. In no place is the fall perpendicular to any considerable extent, but in the distance of half a mile the waters descend about 50 feet. A canal three-fourths of a mile long, with locks, was constructed round the Falls, many years since, at an expense of $50,000.

Keene is one of the prettiest towns of New Hampshire in this vicinity. It

is situated on a flat, east of the Ashuelot river, and is upon the route of the Cheshire railway, by which it is connected with Boston, and with the Connecticut river roads. It is particularly entitled to notice for the extent, width, and uniform level of its streets. The main street, extending one mile in a straight line, is almost a perfect level, and is well ornamented with trees. It is a place of considerable business, there being several manufacturing establishments here.

From Bellows' Falls we pass on to Windsor, 26 miles, by the Sullivan railway.

SOUTH CHARLESTON and CHARLESTON are quiet little aside villages on the east bank of the Connecticut, in Sullivan County, New Hampshire, 50 miles west of Concord. A bridge crosses the river to Springfield, Vermont. Charleston was the extreme northern outpost in the early days of the New England colonies. There was then a rude military work here called Fort No. 4.

CLAREMONT is also on the east bank of the Connecticut, and in Sullivan County, N. H. It is a pleasant little manufacturing village. The scenery in this neighborhood is extremely fine. The banks of the SUGAR RIVER are very picturesque, and the changing aspects of MOUNT ASCUTNEY, which we now approach, are of the highest interest. It is upon this side this noble hill, standing solitary and alone, a brave outpost of the coming Green Mountains on the one hand, and of the White Mountains on the other, is seen in its greatest grandeur. Its rugged precipitous summits and its dark ravines have here a very vigorous and massive character. Ascutney is sometimes called the Three Brothers, from its trio of lofty peaks, all visible from the southern approach. From the eastward and northward, at Windsor and from the west, its appearance is totally different, but always fine. It may be very comfortably ascended from Windsor, in a good day's tramp; and the view from the summit is scarcely inferior in extent, variety, and magnificence, to that from any other peak of the Vermont chain. Its height is 1,732 feet above the river.

WINDSOR is one of the plesantest rural retreats of all this charming region, with its vicinage to Mount Ascutney, and other attractive scenes of land and water. It is the centre of a fine agricultural and wool-growing neighborhood. There is an excellent, quiet, summer hotel here. Windsor is the seat of the Vermont State Prison, and the terminus of the Vermont Central Railway, from Burlington through the valley of the Winooski river.

At Windsor, the Sullivan road ends, and we continue our journey along the Connecticut, 14 miles, to White River Junction, by the Vermont central route.

White River Junction. From this point the Vermont central road continues, via Northfield and Montpelier, to Burlington; and we leave it for the Connecticut and Passumpsic, upon which we continue, 40 miles, to Wells River.

Hanover, New Hampshire, is in this neighborhood, half-a-mile east of the Connecticut, and 55 miles northwest of Concord. It occupies a broad terrace, 180 feet above the water. Here is the venerable Dartmouth College, founded in 1769, and named in honor of William, Earl of Dartmouth. Daniel Webster was one of the alumni of this esteemed institution.

The College buildings are grouped around a square of 12 acres, in the centre of the plain upon which the village stands.

Wells River. At this point, the railway to Littleton, 20 miles, and thence by stage to the White Mountains, diverges. Here, too, comes in the Boston, Concord, and Montreal route, sending its passengers, via Littleton, to the White Hills or onward, by the Connecticut and Passumpsic road, via St. Johnsbury, to Canada. The Connecticut now assumes the appearance of a mountain stream, the railways follow its banks no farther, and we leave our traveller to proceed on either hand, as we have indicated, to New Hampshire, or to Canada.

CONNECTICUT.

THE scenery of Connecticut is delightfully varied by the passage of the Connecticut, the Housatonic, and other picturesque rivers; and of several low hill ranges. Spurs of the Green Mountains rise, here and there, in isolated groups or points through the western portions of the State. The Talcot, or Greenwood's Range, extends from the northern boundary almost to New Haven. Between this chain and that in the extreme west, lies another ridge, with yet two others on the eastward, the Middletown Mountains, and the line across the Connecticut, which is a continuation, most probably, of the White Hills of New Hampshire. Lying between these mountain ranges, are valleys of great luxuriance and beauty. The lakes among the mountains of the north-western corner of the State are extremely attractive.

Excepting a trading house built by the Dutch at Hartford, in 1631, the first colony planted in Connecticut was the settlement of some of the Massachusetts emigrants at Windsor. Soon afterwards Hartford fell into the possession of the English colonists. Wethersfield was next occupied, in 1636, and New Haven, in 1638. The State had its share of Indian troubles in its earlier history, and of endurance, later, in the days of the Revolution.

THE HOUSATONIC VALLEY, RIVER, AND RAILROAD.

The valley of the Housatonic, traversed by the Housatonic river and railroad, extends for about 100 miles northward from Long Island Sound, through the extreme west of Connecticut and Massachusetts, including the famous county of Berkshire in the latter State. The whole region is replete with picturesque and social attractions, and has long been resorted to for summer travel and residence. It is a county of bold hills, pleasant valleys, and beautiful streams—more particularly that portion lying in Berkshire. Saddle Mountain, in the north part of this county, is the highest land in Massachusetts. The natural beauties of Monument Mountain, also in Berkshire, have been heightened by traditionary story, and by the verse of Bryant. Stockbridge and Great Barrington—very popular summer homes—are here. Lenox, honored by the residence of the authoress, Miss C. M. Sedgwick; and Pittsfield, the home of Melville and Holmes. North and South Adams, too, and Williamstown, the seat of Williams College—but we will follow the line of the valley, and glance, briefly, at its points of interest in due order.

From New York, take the New Haven Railroad, 58 miles, to Bridgeport, on the Sound, thence up the valley, on the Housatonic road; or take the Hudson river, or the River Railroad route, 116 miles, to the city of Hudson, and thence by Hudson and Berkshire Railroad, 34 miles, to West Stockbridge; or the Harlem Railroad, to its intersection with the Hudson and Berkshire, at Chatham Four Corners.

From Albany, by the Albany and Boston road, 38 miles, to State Line (Housatonic road); or onward to Pittsfield. From Boston by western (Albany) road, 151 miles, to Pittsfield.

Falls Village, 67 m. above Bridgeport. The Falls here, which are the largest in Connecticut, are very bold and picturesque. The waters traverse a ledge of limestone, and make a descent of 60 feet.

The Salisbury Lakes. The country west of Canaan, as all this part of the State, is beautifully embellished with hill and lake scenery. The Twin Lakes, in Salisbury, are very charming waters.

Sheffield is a prosperous village, famous for its manufactures, and for its varied attractions in hills and cascades, and other forms of natural beauty.

Great Barrington, with excellent hotels for summer travel, is a place of favorite resort. Mount Peter, on the southern edge, overlooks the village pleasantly; and it is most agreeably seen approaching, on the river road, from the north.

The Taughkanic Mountains, a range extending from the Green Hills of Vermont, lie between the Housatonic valley and the Hudson River. Mount Washington, Mount Riga, and other peaks, are interesting places of pilgrimage and exploration. The Falls of Bashpish are in this hill range. Following the Housatonic, and passing Monument Mountain, we reach

Old Stockbridge. This is one of the quietest and most winsome retreats in the world, lying in the lap of a fertile, hill-sheltered valley. The houses, which are all far apart, and buried in dense verdure, stand back in gardens, upon either side of a broad street or road, thickly lined with noble specimens of the ever-attractive New England elm. There is a pleasant, well-ordered hotel here. Miss Sedgwick has, in her stories, woven much romantic interest about many spots in this vicinity, and about her own home of Lenox near by.

Lebanon Springs (N. Y.), and the Shaker village are hereabouts. (See New York.)

Pittsfield, Berkshire County, Mass., is a large manufacturing and agricultural town, elevated 1,000 feet above the level of the sea. It is 151 miles west from Boston, and 49 east from Albany. The village is beautifully situated, and contains many elegant public edifices and private dwellings. In this village there is still standing one of the original forest trees—a large elm, 120 feet high, and 90 feet to the lowest limb—an interesting relic of the primitive woods, and justly esteemed a curiosity by persons visiting this place. The town received its present name in

Valley of the Housntonic.

Monument Mountain, Mass.

1761, in honor of William Pitt, (Earl of Chatham).

Upon a fine spacious square in the heart of the town, are the principal hotels, the Berkshire Medical School, a popular institution, founded in 1823, and the First Congregational Church, a Gothic structure of stone, erected in 1853. There is, too, a prosperous Young Ladies' Institute here, which occupies several admirable buildings, surrounded by well-embellished grounds. Pittsfield is a large depot of manufactures, being extensively engaged in the production of cotton and woollen goods, machinery, fire-arms, and railroad cars. The population of the township is nearly 7,000. It is upon the Western Railway, from Boston to Albany, at the northern terminus of the Housatonic valley route, and at the southern terminus of the Pittsfield and North Adams Railway.

The scenery of this region, traversed by the western road through Berkshire, from Boston to Albany, is often of very impressive aspect.

After leaving the wide meadows of the Connecticut, basking in their rich inheritance of alluvial soil, and unim-peded sunshine, you wind through the narrow valleys of the Westfield river, with masses of mountains before you, and woodland heights crowding in upon you, so that, at every puff of the engine, the passage visibly contracts. The Alpine character of the river strikes you. The huge stones in its wide channel, which have been torn up and rolled down by the sweeping torrents of spring and autumn, lie bared and whitening in the summer sun. You cross and recross it, as in its deviations it leaves space, on one side or the other, for a practicable road.

At "Chester Factories" you begin your ascent of 80 feet in a mile for 13 miles. The stream between you and the precipitous hill side, cramped into its rocky bed, is the Pontoosne, one of the tributaries of the Westfield river. As you trace this stream to its mountain home, it dashes along beside you with the recklessness of childhood. It leaps down precipices, runs forth laughing in the dimpling sunshine, and then, shy as the mountain nymph, it dodges behind a knotty copse of evergreens. In approaching the "summit level," you travel bridges built 100 feet above other mountain streams, tearing along their deep-worn beds; and at the "deep cut" your passage is hewn through solid rocks, whose mighty walls frown over you.

The Pittsfield and North Adams Railroad Route. This road extends 20 miles, via Packard's, Berkshire, Cheshire, Cheshire Harbor, Maple Grove, and South Adams to North Adams.

Adams. The villages of North and South Adams are in the immediate neighborhood of Saddle Mountain. This noble peak has an elevation of 3,500 feet, and is the highest land in Massachusetts. There is a notable natural bridge upon Hudson's brook in this township.

Williamstown, near North Adams, is the seat of Williams College, founded in 1793. This institution is well endowed, and holds high rank among the best educational establishments of the country. The village is in one of the most picturesque portions of picturesque Berkshire.

New Haven, Hartford, &c. For mention of these and other cities and scenes of Connecticut, see Index and Routes to Boston from New York, under the head of Massachusetts.

RHODE ISLAND.

RHODE ISLAND is the smallest of the many States of the great American confederacy, her entire area not exceeding 1306 miles, with an extreme length and breadth, respectively, of 47 and 37 miles.

The country is most pleasantly varied with hill and dale, though there are no mountains of any great pretensions. Ample compensation for this lack in the natural scenery, is made by the numberless small lakes which abound everywhere, and especially by the beautiful waters and islands and shores of the Narragansett Bay, which occupy a great portion of the little area of the State. Its capitals, Providence and Newport, are among the most ancient and most interesting places in the land, and the latter has of late years become the most fashionable of all the numerous American watering-places.

Rhode Island was first settled at Providence, in 1636, by Roger Williams. To the enlightened and liberal mind of Williams in Rhode Island, and to the like true wisdom of Penn in Pennsylvania, and of Lord Baltimore in Maryland, America owes its present happy condition of entire freedom of conscience;—perfect religious toleration having been made a cardinal point in the policy of these colonies.

The people of Rhode Island were early and active participants in the war of the Revolution, and many spots within her borders tell thrilling tales of the stirring incidents of those memorable days.

PROVIDENCE AND VICINITY.

Providence, one of the most beautiful cities of New England, and surpassed only by Boston in wealth and population, is a semi-capital of Rhode Island, on the northern arm of the Narragansett Bay, called Providence River. It is an ancient town, dating back as far as 1635,— when its founder, Roger Williams, driven from the domains of Massachusetts, sought here, that religious liberty which was denied to him elsewhere. It bears its venerable age, however, bravely, and looks, to-day, as youthful and vigorous as the Aladdin cities of yesterday—yet with the accumulated refinements and amenities, in its social character, of very many cultivated generations. This city makes a charming picture seen from the approach by the beautiful waters of the Narragansett, which it encircles on the north by its business quarter, rising beyond and rather abruptly, to a lofty terrace, where the quiet and gratefully shaded streets are filled with dainty cottages and grand manorial homes. Providence was once a very important commercial depot, its rich ships crossing all seas—and at the present day the city is equally distinguished for its manufacturing wealth and enterprise. In this department of human achievement it took the lead, which it still keeps,—the first cotton mill which was built in America, being still in use, in its suburban village of Pawtucket, and some of the heaviest mills and print-works of the Union being now in operation within its borders. The value of the annual product of the cotton mills and print-works of Providence, is estimated at nearly four millions of dollars; that of the manufacturers of jewelry of various kinds—its establishments in this labor being no less than from sixty to seventy in number—at two and a half millions. It has also extensive manufactories of steam machinery, and of tools and implements of all sorts, and it furnishes the major part of all the screws used in the United States. The workshops of the Eagle Screw Company, where these little implements are made, are among the best appointed in the world. The total capital invested here, in manufactures, is six millions of dollars.

Providence is the seat of BROWN UNIVERSITY, one of the best educational establishments in America, founded in Warren, R. I., 1764, and removed to Providence in 1770. Its library is very large, valuable, and is remarkably rich in rare and costly works.

The ATHENÆUM has a fine reading-room, and a collection of 12,000 books. The BUTLER HOSPITAL for the Insane, upon the banks of the Seekonk River, is an admirable institution, occupying large and imposing edifices. In the same part of the city, and lying also upon the Seekonk River, is the Cemetery, a spot of great rural beauty. There are about fifty public schools in Provi-

dence, in which instruction is given to between six and seven thousand pupils, at an annual expense of over $45,000, one-fourth only of which is contributed by the State. The Dexter Asylum for the Poor, is upon an elevated range of land east of the river. In the same vicinage is the yearly meeting boarding school, belonging to the Society of Friends. The Reform School occupies the large mansion in the south-east part of the city, formerly known as the Tockwotton House. The Exchange, the Rail Road Depot, some of the banks, and many of the churches of Providence are imposing structures.

The topography and the natural scenery of Providence and its vicinity, are great temptations to tourists and to those seeking pleasant summer abodes. Situated upon the shore of the Narragansett Bay, and connected with it at all points by railway and steamboat, it unites all the pleasures of city and country life.

Upon the immediate edge of the city, on the shore of a charming bay in the Seekonk River, is the famous WHAT CHEER ROCK, where the founder of the city, Roger Williams, landed from the Massachusetts side to make the first settlement here. From this rock he was greeted by the Indians, with the salutation which gives name to the spot.

At HUNT'S MILL, three or four miles distant, is a beautiful brook with a delicious little cascade, a drive to which is among the morning or evening pleasures of the Providence people and their guests.

VUE DE L'EAU, is the name of a picturesque and spacious summer hotel, perched (four or five miles below the city) upon a high terrace overlooking the Bay and its beauties for many miles around.

Gaspee Point, below, upon the opposite shore of the Narragansett, tells a stirring story of the initial days of the Revolution, when some citizens of Providence, after adroitly beguiling an obnoxious British revenue craft upon the treacherous bar, stole down by boats in the night and settled her busi-

Places and Scenes in the Narragansett Bay.

ness by burning her to the water's edge.

Rocky Point, is a wonderful summer retreat among shady groves and rocky glens, upon the west shore of the Bay. In summer-time half a dozen boats ply, each twice a day, on excursion trips from Providence to various rural points down the Bay, charging 25 cents only for the round voyage.

Rocky Point is the most favored of all these rural recesses. Hundreds come here daily and feast upon delicious clams, just drawn from the water and roasted on the shore, in heated seaweed, upon true and orthodox "clam bake" principles. Let no visitor to Providence fail to eat clams and chowder at Rocky Point, even if he should never eat again.

The charming towns of Warren and Bristol are across the Bay, each worthy of a long visit. They may both be reached several times a day from Providence, by the Providence, Warren and Bristol Railroad.

Mount Hope, the famous home of the renowned King Philip, the last of the Wampanoags, is just below Bristol, upon Mount Hope Bay, an arm of the Narragansett on the east. The bare crown of this picturesque height presents a glorious panorama of the beautiful Rhode Island waters. Upon the shore of Mount Hope Bay, opposite, is the busy village of Fall River, which we have already visited, on our route to Boston from New York. Off on our right, as we still descend towards the sea, is Greenwich, and near by it, the birthplace and home of General Nathaniel Greene—the revolutionary hero—and just below is the township and (lying inland) the village of Kingston. In this neighborhood there once stood the old snuff-mill in which Gilbert Stuart, the famous American painter, was born.

Prescott's Head Quarters, is a spot of Revolutionary interest, on the western shore of the large island filling lower part of the Bay, the island after which the State is named.

At the southern extremity of the island, is the venerable town of Newport, at this day the most fashionable of all summer watering-places. Leaving Newport for a chapter by itself, let us, now that we have run rapidly down the 35 miles of the Narragansett waters,

Mount Hope, Rhode Island.

Fort Dumplings, Newport, R. I.

return for another moment to Providence.

We may get here any day from New York, by the Fall River route for Boston, round Point Judith from the sea, up the Narragansett (calling at Newport) to Fall River, in Mount Hope Bay, and thence by a Providence steamer. Or we may come by the Stonington route from New York, to Stonington by steamer, thence by rail, or more speedily, morning or afternoon, by the charming route along the marge of Long Island Sound—the new "Shore Line" via New Haven, New London, etc. ; or we may come from Boston, any hour, almost, by the Boston and Providence road.

Distance from New York to Providence, about 175 miles—usual fare, $3 to $4. Distance from Boston, 43 miles —fare, $1 50. Population of Providence, about 50,000.

Hotels. Providence is a notable exception to the general rule of American .cities, in respect to the provision of public hospitality. The only house which we can commend to the stranger is the City Hotel, on Broad street.

NEWPORT.

If Newport were not, as it is, the most elegant and fashionable of all American watering-places, its topographical beauties, its ancient commercial importance, and its many interesting historical associations, would yet claim for it distinguish-ed mention in these pages. Coming in from the sea round Point Judith, a few miles bring the traveller into the waters of the Narragansett Bay, where he passes between Fort Wolcott, on Goat Island, and the stronghold of **Fort Adams,** upon Brenton Point on the right, and enters the harbor of the ancient town, once among the commercial capitals of the Union.

In the Revolution, the British long held possession of Newport, during which time, and at their departure, it became almost desolate. Before leaving, they destroyed 480 buildings, burned the lighthouse, cut down all the ornamental and fruit trees, broke up the wharves, used the churches for riding-schools, and the State House for an hospital, and carried off the church bells and the town records to New York; disasters which reduced the population from 12,000 to 4,000. But the incidents of this period have left some pleasant memories for the present day ; and remembrances of the fame of Commodore Perry, the gallant commander on Lake Erie, who was born in Narragansett, R.I., across the bay, and whose remains lie now in Newport ; of the residence of Rochambeau, and other brave officers of the French fleet ; and of the visits of General Washington, and the *fêtes* given in his honor ; the venerable buildings associated with all these incidents being still to be seen.

The Old Stone Mill, Newport.

The old town lies near the water, but of late years, since the place has grown so great as a summer residence, a new city of charming villas and sumptuous mansions has sprung up, extending far along upon the terraces which overlook the sea. The flood of travel has called up, too, a number of magnificent hotels, of which the chief are, the Ocean House, charmingly situated on Touro street; the Atlantic House, at the head of Pelham street; the Bellevue House, on Catherine street; the Aquid-neck and the Fillmore. Of the old buildings, and of those which belong to Newport *per se*, instead of in its character of a watering-place, are, the ancient State House (for Newport is a semi-capital of Rhode Island), the Redwood Library and Athenæum, the Old Stone Mill, an interesting relic of a period past remembrance and almost of tradition; Tammany Hall Institute, Trinity Church, the Vernon family mansion, the Perry monument, Com. Perry's house, the City Hall, the fortifications in the harbor, Fort Adams, Fort Wolcott, Fort Brown, and its surrounding rocks called the *Dumplings*, etc.

The chief picturesque attractions of the town and its immediate vicinity, are the fine ocean shores, known as the First, the Second, and the Third Beach. It is the First Beach which is chiefly used as a bathing ground by the Newport guests. At the Second Beach are the famous rocks called Purgatory, and the Hanging Rocks, within whose shadow it is said that the celebrated Bishop Berkeley wrote his "Minute Philosopher." Here, too, groups of fishermen may be seen drawing their nets for the manhaden and blue-fish, or

Bluff near Purgatory, Newport, R. I.

horse-mackerel, with which the waters abound. Nearer to the town, and upon the coast, is the great Spouting Cave.

The Glen, and other localities, are charming places to ride to, when the weather is auspicious.

Newport was the birthplace of the gifted miniature painter Malbone, and Gilbert Stuart's place of nativity may be seen in Narragansett, across the bay. Stuart made two copies of his great Washington picture for Rhode Island, one of which may be seen in the State House at Newport, and the other in that at Providence, the twin capital.

A steamboat passes every day up and down the bay between Newport and the city of Providence, enabling the traveller to see at his leisure the many sights and localities of all this attractive neighborhood. See routes from New York to Boston for ways of approach to Newport, and chapter upon Providence and vicinity for scenes upon the Narragansett waters.

NEW HAMPSHIRE.

NEW HAMPSHIRE contains some of the grandest hill and valley and lake scenery in America. The White Mountains here are popularly supposed to be the highest land east of the Mississippi river, as, indeed, they are, with the single exception of Black Mountain in North Carolina. These noble hills occupy, with their many outposts, a very considerable portion of the State, and form the speciality in its physical character. The reader will find a detailed mention of all these features, and of the beautiful intermediate lake-region, in subsequent pages.

On his route from Boston to the mountain regions, the tourist will find much to interest him, if his interest lies that way, in the enterprising manufacturing towns of the lower part of the State. In its historical records, New Hampshire has no very striking passages—no important reminiscences, either of the Revolutionary War or of the later conflict with Great Britain in 1812.

The railway lines of New Hampshire are numerous enough to give ready access to all sections of her territory, and to the neighboring States. Occasion will occur for ample mention of the facilities which they afford for travel, as we follow them, severally, hither and thither.

ROUTES TO THE WHITE MOUNTAINS, N. H.

FROM NEW YORK, BOSTON, PORTLAND, ETC.

Via Boston.

ROUTE 1. From Boston, by Lake Winnipiseogee and Conway Valley. (See routes from New York to Boston.) From Haymarket Square, Boston, at 7 30 A. M. and 12 M., 26 miles to Lawrence by Boston and Maine road; 27 miles to Manchester, upon Manchester and Lawrence R. R.; 18 miles to Concord, upon Concord R. R.; 33 miles to Weirs, on Lake Winnipiseogee, by Boston and Concord and Manchester R. R.; 10 miles by steamer Lady of the Lake, on Lake Winnipiseogee to Centre Harbor (dine at Centre Harbor); 30 miles by stage to Conway; arrive at Conway in the evening, remain there all night, and proceed, 24 miles, to Crawford House, White Mountain Notch, next day. Total distance from Boston to the Crawford House, 168 miles; time, 2 days and 1 night; fare, $7 45. Distance from New York, 431 miles; time, 2 days and 2 nights; fare, about $12 45. Passengers by the Boston morning train only reach Conway the same evening. Those taking No. 2, or noon train, will pass the night at Centre Harbor, on Lake Winnipiseogee, and

hornton's Ferry

41 m.ᵉ from
& 42 fr. Worces

utles
7

Pelham
1

Tynas Long P.

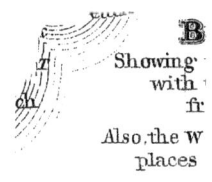

B
Showing
with
fr
Also the W
places

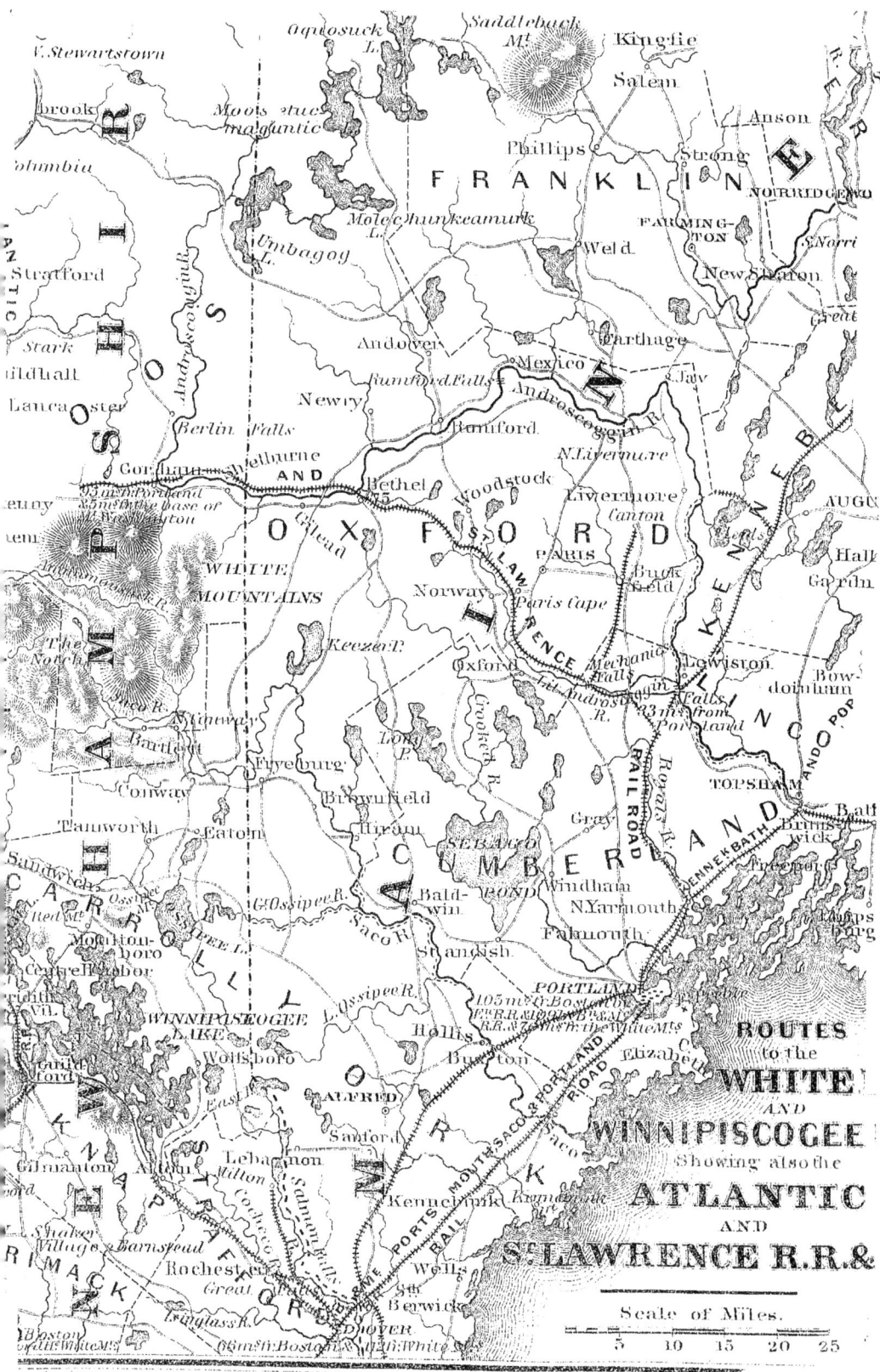

ROUTES
to the
WHITE
AND
WINNIPISCOGEE
Showing also the
ATLANTIC
AND
St. LAWRENCE R.R. &

Scale of Miles.
5   10   15   20   25

Routes to the White Mountains.

the next night at Conway, reaching the mountains on the third day.

ROUTE 2. Fom Boston. (See routes from New York to Boston.) Leave Haymarket Square (as in route 1) at 7 30 A. M. and 12 M.; 68 miles to Dover, N. H., upon Boston and Maine R. R.; thence to Alton Bay, 28 miles, upon Cocheco R. R.; thence, 30 miles, by steamer Dover (dine on board) to Wolfboro' and Centre Harbor, on Lake Winnipiseogee; thence by stage, via Conway, to the mountains, as in Route No. 1.

Passengers by morning train only, from Boston, reach Conway same night. Those by second, or noon train, will pass the night at Wolfboro' or Centre Harbor. From Boston to Crawford House, by this route, 96 miles by railroad, 30 by steamboat, and 54 by stage; total, 180. Time, 2 days and 1 night, from Boston; fare, $7 45.

ROUTE 3. From Boston. At 7.30 A.M. and 12 M. (station, Causeway street), 26 miles to Lowell, by Boston and Lowell R. R.; 15 miles to Nashua, upon Nashua and Lowell R. R.; 35 miles to Concord, upon Concord R. R.; afterwards as in Route No. 1. Distance, time, and fare the same.

ROUTE 4. From Boston, same as in Routes Nos. 1 and 3, as far as Weirs on Lake Winnipiseogee; thence, continuing upon the railroad, 18 miles from Weirs, to Plymouth, N. H; dine at Plymouth, and proceed by stage, 24 miles, through West Campton, etc., to the Flume House, Franconia Notch, the western end of the mountains. Passengers by the morning train from Boston will reach the Flume House, Franconia Notch, same evening. Those taking the second train will stay over until next day at Plymouth. Distance from Boston to Flume House, 148 miles, being 124 by railway and 24 by stage. Time, from Boston by morning line, 10 hours; fare, $5 15. Stages daily from Flume House, 5 miles, to the Profile House, 22 miles to White Mountain House; thence, 5 miles, to Crawford House, terminus of Routes 1, 2, and 3, on the east side. Distance from Flume House

to Crawford House, 32 miles. Fare, $3.

ROUTE 5. From Boston, same as in Routes 1 and 3, to Weirs; thence (as in Route 4) to Plymouth (dine), continuing upon the railroad, 42 miles, from Plymouth to Wells River; thence upon White Mountain R. R., 20 miles, to Littleton; thence by stage, 11 miles, to Profile House, and 5 miles further to Flume House, or 23 miles to Crawford House. Passengers by the early train only reach the mountains the same night. Those taking second train stay till next day at Plymouth. From Boston to Profile House, 193 miles; to Flume House, 198 miles; to Crawford House, 205 miles. 182 miles by railroad, rest by stage. Fare from Boston to Profile House, $6 15; to Crawford, $6 90. Time, 12 hours.

From New York or Boston, via Portland, Maine.

ROUTE 6. (See routes from New York to Boston.) Leave Boston for Portland, 111 miles by railway, morning and evening, from Haymarket street, via Reading, Lawrence, Haverhill, Exeter, etc. Through baggage for the White Mountains to be marked, "*Portland East.*" Passengers by first train will dine in Portland, and take Grand Trunk Railway through Cumberland, Yarmouth, etc., 91 miles, to Gorham, N.H. Second train passengers will pass the night at Portland, and proceed to Gorham next day. From Gorham, 8 miles, by stage to Glen House, foot of Mount Washington. Stages leave Glen House every morning for Crawford House, 34 miles distant, via Pinkham Notch, also via Cherry Mountain. From Boston to Gorham, 202 miles; from Boston, via Portland, Gorham, Glen House, and Pinkham Notch, to Crawford House, 244 miles.

ROUTE 7. From Boston, via Portland. Leave Causeway street, morning and evening, by railway, through Lynn, Salem, Beverly, Newburyport, Portsmouth, etc., to Portland, and thence as in Route No. 6.

ROUTE 8. From Boston by steamer to

Portland, every night, from end of Central Wharf; thence, as in Route No. 6. Fare by this line, $3, from Boston to Gorham.

ROUTE 9. From Boston to Portland, by railway or steamer, as in Routes 6, 7, and 8, and thence by Sabago Lake and Pleasant Mountain to Conway; thence to Crawford House, etc., as in Route 1.

From New York, NOT *via Boston.*

ROUTE 10. From New York by railway, via New Haven, Hartford, and Springfield; thence by railway up the Valley of the Connecticut to Wells River, and from thence to Littleton, N. H.; from Littleton by stage, as in route 5.

ROUTE 11. From Pier 18, Courtlandt street, North River, N. Y., every evening to Allyn's Point; thence by railway to Worcester, Nashua, and Concord; and from Concord on the east side by Conway to Crawford House, route 1; or on the west side by Campton to the Franconia Notch, route 5.

ROUTE 12. From New York by Hudson River, or Hudson River Railway, to Albany and Troy; thence to Whitehall, and down Lake Champlain to Burlington, Vermont; thence by Vermont Central Railroad through the Winooski Valley and Green Mountains (via Montpelier), to connections with the Connecticut Valley roads to Littleton, N. H.

ROUTE 13. From New York by Hudson River to Albany; thence to Whitehall, foot of Lake Champlain, or other routes to Rutland, Vermont; thence to Bellows' Falls, on the line of the Connecticut Valley road, to Littleton, N. H.

We might much extend our list, but as all roads lead to Rome, so the ways to the favorite summer haunts in the White Mountains are infinite.

DESCRIPTION OF ROUTES TO WHITE MOUNTAINS.

ROUTE 1. By Lake Winnipisseogee and Conway Valley. From Boston, 26 miles, to Lawrence, via Boston and Maine Railroad (Boston and Portland), passing numerous suburban villages, for which see "Boston and Vicinity." Lawrence is a large manufacturing city (population 15,000), upon the Merrimac River. It is connected with Lowell (13 miles distant), with Salem, 21 miles, and with all surrounding points by railway. It has grown up suddenly within a few years, having been incorporated only in 1845.

From Lawrence by Manchester and Lawrence Railroad, 26 miles to Manchester, N. H., still following the Merrimac River, upon which Manchester, like Lawrence, is situated. At this point we are 53 miles above Boston, and 18 miles below Concord. This place has, like Lawrence and others, suddenly grown under the development of manufacturing enterprise—from an inconsiderable village, into a large and populous city. Its charter was granted in 1846, and in 1853 it had 20,000 inhabitants.

Concord, the capital of New Hampshire, is upon the banks of the Merrimac, 18 miles above Manchester, by the Concord Railroad. The State Capitol, the Lunatic Asylum, the State Prison, are public edifices of interest. A Methodist General Biblical Institute was founded here in 1847. We might suggest to the tourist a brief halt at Concord, were he not now so near yet more attractive scenes. Concord is 71 miles from Boston, via Lawrence; 47 miles from Portsmouth, N. H., by Concord and Portsmouth Railroad; 25 miles from Bradford, by the Concord and Claremont Railroad; 35 miles from Nashua, by Concord Railroad; 93 miles from Wells River, by Boston, Concord, and Montreal Railroad. Concord is an eligible place for summer abode. It is the home of ex-President Pierce.

From Concord, our route follows the Boston, Concord, and Montreal Railroad, 33 miles to Weirs, on Lake Winnipiscogee, where we take the steamer Lady of the Lake, 10 miles to Centre Harbor. Our White Mountain route, No. 5, continues on this road, past

WOODSTOCK Hartford
70 miles fr. Concord
Hartland
WINDSOR
Cavendish
W. Clarendon
Ludlow
Wethersfield
Springfield
Chester
Saxtons River
BELLOWS
FALLS 65 m fr. Green
field 18 fr. Keene &
110 fr. Boston
West-
minster
Putney
Marlboro'
TLEBOROUGH
reenfield 69 fr.
19 fr. Boston
rd
tre
olernine
Bernardston
Gill
NS
REE
FIELD
ring
stown
Deerfield

Plainfield
Cornish
Newport
L. Sunapee
Claremont
Newbury
Charlestown
Walpole
Sullivan
West
moreland
KEENE
42 m fr. Fitchburg
& 92 fr. Boston
Chester-
field
Swan-
zey
Troy
Winchester
Vernon
Fitzwilliam
Northfield 50
m fr. Fitchburg
MASS. R.R.
Miller's R.
Wendell
Royalston
Athol
Philips-
ton

Canaan 52
Newfound
GRAFTON 44
field 59
BRISTOL
Danbury
W. Andover 3
Andover
E. Andov.
Springfield
L. Sunapee
MERRIMACK
Bradford
Warner
Henniker
Windsor
Hillsboro'
Stoddard
Deering
Hancock
Francistown
N. Boston
Marlboro'
Peterboro'
Jaffrey
Wilton
New Ipswich
Mason
Rindge
Brookline
Winchenden
Ashby
Royals-
ton
Ashburnham
Baldwins-
ville
Gardner
Templeton

Meridith Hill
Part of Lake
Winni-
piseogee
Sanborn-
ton
Gilford
Union
Franklin
18 m from Concord
N. Boscawen 15
Boscawen 10
Salisbury
Canterbury
Fishersville 7
CONCORD
35 m fr. Nashua 876 from
Boston
Pem-
broke
Hopkinton
Bow
Dunbarton
Goffstown
Amoskeag
Bedford
15 miles from
Nashua
AMHERST
Southegan
Merrimack
NASHUA
Hollis
MIDDLE
Town
send
FITCHBURG
50 m from
Boston

Weirs, to Wells River, and Littleton, N. H.

Lake Winnipiseogee. The little voyage on this beautiful lake, is among the most agreeable passages in our present journey to the White Mountains, and well deserves a pilgrimage to itself alone. Winnipiseogee is an enchanting reach of pure, translucent waters, very irregular in form; some 23 miles long, and from one to ten miles wide. It is crowded with exquisite island groups, indented with surprising bays; and bold mountain peaks cast their shadows every where into its still, deep floods.

Red Mountain, about 1,600 feet high, a remarkably beautiful eminence, is situated on the N.W. of the above lake. The ascent to the summit, although steep and arduous, can be effected, for a portion of the distance in carriages and all the way on horseback. From the S.E. there is a fine panoramic view of the lake and the adjacent country. On the south ascends Mount Major, a ridge of a bolder aspect and loftier height. On the N.E. the great Ossipee raises its chain of elevations, with a bold sublimity, and looking down in conscious pride upon the regions below.

Squam Lake lying W. from Red Mountain; and two miles N. W. from Winnipiseogee Lake, is another splendid sheet of water. It is about six miles in length, and in its widest part not less than three miles in breadth, and, like its neighbor (Winnipiseogee), is studded with a succession of romantic islands. This lake abounds in trout of the finest kind.

Centre Harbor, with its excellent summer hotel upon the margin of Winnipiseogee, is the halting place for the explorer of the many beauties of this region. White Mountain tourists dine here *in transitu*, and proceed for the rest of the way by stage coach, first for 30 miles through a country of picturesque delights to Conway Valley, where they might well linger till their summer days all went by.

Conway Valley is a wide stretch of delicious interval lands upon the Saco River, hemmed in upon all sides by bold, mountain summits, chief among which are the stern cliffs of Mount Washington itself. It is a delightful

Concord, New Hampshire.

Lake Winnipiseogee, N. H.

place for artistic study, and for summer residences; and within a few years past, it has been a favorite resort of the American Landscapists, and has grown to be a veritable "watering place," in the great numbers of tourists who not only pass, but linger within its borders. Pleasant hotels and boarding-houses are springing up, and country villas even are beginning to dot its knolls, and to lurk in its verdant glens. The picturesque portion of this valley, *par excellence*, is North Conway, where the Kearsarge House (Thompson's), or the Washington House, offers all desirable hotel appliances. Beside the distant views of the White Mountain ranges, proper, which are of surpassing interest here, Conway is full of local and neighboring attractions of the greatest beauty, as are the broad meadows, and the wooded, winding banks of the Saco; the nooks and turns of the Artists' Brook, and other elfish waters; the Paquawket Mountain, those grand perpendicular cliffs, 650 and 950 feet in height, called the Ledges; the magnificent peaks of Kearsarge and Chicorua; the

Echo Lake, the Crystal Falls, and Diana's Bath.

Conway village and Conway corners are a few miles below North Conway. They are most agreeable places, *en route*, amply supplied with hotel accommodations. Leaving Conway, as the tourist does, the morning following that of his departure from Boston, he continues on through valley and over hill, 24 miles to the Crawford House, where we shall meet him when we have followed over other routes to the threshold of the mountains. We will, however, accompany him yet on his journey from Conway, through Bartlett and Jackson, by the Old Crawford House; and by the famous Willey House, the scene of the awful avalanches of 1826, when the entire Willey family were destroyed. (*See further mention later.*)

ROUTE 2. From Boston, 68 miles via Lawrence to Dover, N. H., on the Boston and Maine Railroad. Dover is a pleasant town of some 8,000 people; upon the banks and at the falls of the Cocheco River, a tributary of the Pis-

White Mountain Routes.

cataqua. Our route leads hence by the Cocheco Railroad to Alton Bay; southern extremity of Lake Winnipiseogee. Here, we take the steamer Dover for Centre Harbor, traversing the entire length of the lake, and proceed thence via Conway, as in Route 1.

ROUTE 3. From Boston, 26 miles, to the famous manufacturing city of Lowell. (See Boston and Vicinity.) From Lowell, 15 miles, to Nashua—an important manufacturing town, at the confluence of the Nashua with the Merrimac river; thence, 35 miles, to Concord, N. H., and from Concord to Weirs and Centre Harbor, on Lake Winnipiscogee, and on, via Conway, as in Routes 1 and 2.

ROUTE 4. From Boston, as in Route 1, or 3, to Weirs, on Lake Winnipiseogee, thence on, without stopping, to Plymouth, N. H., where passengers dine and take stage for the rest of the way; or where they remain all night, if they leave Boston by the noon, instead of the morning train. Plymouth is in the midst of a noble mountain landscape, being the extreme southern threshold of the Franconia range of the White Hills. It is upon the banks of the beautiful Pemigewasset river, near its confluence with Baker's river. The Pemigewasset House here, at the railway station, is an inviting place for summer tarry. The Wells River and Littleton route from Boston to the mountains by the west passes Plymouth.

Leaving Plymouth in the stage, after dinner, we reach the Flume House, at the Franconia Notch, 24 miles distant, the same evening, unless we stop by the way, as would be very reasonable—for the whole journey is through most inviting spots and places. The villages on the route are but little affairs; and there is not a fashionable hotel in a'l the distance, until we reach the Flume; but there are numerous small inns, where artists and their families are well content to pass the summer. There is such an one at

West Campton, a little hamlet on Pemigewasset river, seven miles above Plymouth. West Campton is becoming a greater resort of the landscape painters than North Conway, on the south-east slope of the mountains, has been for several years past. Other tourists will follow, and hotels and boarding-houses will grow up with the demand. The views here, of the Franconia Hills, are especially fine, and the river and brook landscape, with its wealth and variety of vegetation, is of extraordinary interest. The Pemigewasset river, which rises in the little lakes of the Franconia Mountains, winds through all the wonderful valley which we traverse between Plymouth and the Flume House. We shall rejoin our tourist, by and by, at the Flume.

ROUTE 5. To Boston, same as in Route 1, to Weirs on Lake Winnipiseogee, thence on, without halt (as in Route 4), to Plymouth, N. H. Thence, after dinner (*morning* train from Boston), still upon the railroad, 42 miles, to Wells River, Vermont.

Wells River is at the junction of the Connecticut and Wells river—a famous meeting-point of railway lines. Our present route meets here with the Connecticut valley road to the White Mountains. The Vermont Central Road, from Burlington, on Lake Champlain, comes in at White River Junction, 40 miles below. From Wells River our route proceeds by White Mountains Railroad, 20 miles, to Littleton, and for the rest of the way, by stage, either to the Franconia Notch, 12 miles (Profile House), or to the Eastern or White Mountain Notch (Crawford House), 22 miles.

ROUTE 6. Via Portland, and through Maine, on the east side of the mountains. This route, as Routes 7, 8, and 9, are all agreeable approaches to the White Hills, but more circuitous from New York or Boston, than either of the Routes 1 to 5. The Boston and Maine, one of the two railways from Boston to Portland, runs (111 miles) east of north and always near the Atlantic coast, through portions of Massachusetts, and through New Hampshire. (See "Boston to Portland.") From Portland, our present route is by the Grand Trunk Railway, 91 miles, to Gorham, N. H. The Rev. Mr. Willey, in his "Incidents in White Mountain History," says of Gorham, that "it is a rough, unproductive township, lying on the northerly base of the mountains. The opening of the Atlantic and St. Lawrence Railway (the Grand Trunk) brought the little town out from the greatest obscurity, and it has become one of the favorite resorts for the travelling community. Its peculiarly favorable situation for viewing the mountains was never known until travellers, posting through its borders, for other destinations, were compelled to admire its beauties.

"Immediately on the completion of the railroad to this point, the Alpine House was erected, and the announcement made, that the cars set passengers down at the very base of the White Mountains. People, for a moment, were dumb with astonishment. It had never been supposed that there was any north or south, or east or west, to these old heights; but that every one who visited them must make up his mind for a long stage-coach ride through Conway or Littleton, and ultimately be set down at the Crawford or Fabyan's. That the cars should actually carry visitors to the base of the mountains was something which every one had supposed would take place in the far-off future, but not until they themselves had ceased to travel; but it was certainly so; and the Alpine House and Gorham had become familiar words to travellers.

"The Alpine House is a large hotel, owned by the railroad company. It is some distance from the base of the mountains, which are seldom ascended from this point; but for quiet and comfort, and beautiful drives, it is surpassed by no house in the White Hills. A beautiful little village has sprung up around it, consisting mostly of buildings owned by the company. The Post Office is kept here, and the telegraph affords an excellent opportunity to business men to visit the mountains and attend to their business at the same time. Mount Moriah, Randolph Hill, Berlin Falls, and Lary's, should all be visited before the traveller takes his departure."

The Glen House, our next point, (seven miles from Gorham), is, says Mr. Willey, whom we have just quoted, "in the valley of the Peabody river, immediately under Mount Washington, and in the midst of the loftiest summits in the whole mountain district. The house is situated in Bellows' clearing, which contains about 100 acres. For a base view of the mountains, no spot could be selected so good. Several huge mountains show themselves proudly to view, in front of the piazza, nothing intervening to obscure their giant forms. You see them before you in all their noble, calm, and silent grandeur, severally seeming the repose of power and strength. On the left is the *mountain* bearing the *worthiest name* our country ever gave us. Toward the right of its rock-crowned summit rise, in full view, the celebrated peaks of Adams and Jefferson—the one pointed, the other rounded. On both wings of these towering summits are the tops of lesser elevations. In an opposite direction, fronting the 'patriot group' of gigantic forms, is the long, irregular rise of Carter Mountains."

It is from the Glen House that a famous carriage way was to lead to the summit of Mount Washington. This road was to be eight miles long, and was to be made to the very crown of the lofty mountain; 15 feet wide, clear of all obstructions, and Macadamized throughout. The average grade was to be an ascent of one foot to eight and a half feet, with frequent stations at eligible points of view. The estimated cost of this road was 100,000 dollars. A magnificent hotel was to be built upon the mountain top. This project, after four miles of road had been built, and a portion of it macadamized, was suspended, and thus remains.

From the Glen House we must now reach the Crawford House, on the mountain, 34 miles distant, via the Pinckham Notch, or by Cherry Mountain.

Mount Washington, N. H.

Route 7, is from Boston to Portland, by the Eastern (the lower) Railway, through Lynn, Newburyport, and other towns in Massachusetts, and Portsmouth, in New Hampshire. From Portland we proceed by Grand Trunk road to Gorham, as in Route 6.

Route 8. From Boston to Portland, by steamer, and thence as in Route 6.

THE WHITE MOUNTAINS—SCENES AND INCIDENTS.

These mountains cover an area of about 40 miles square, in Northern New Hampshire; though the name of White Mountains is, in the neighborhood, given to the central group only—the half-dozen lofty peaks, of which Mount Washington is the royal head and front. The western cluster is contra-distinguished as the Franconia range. We will suppose our tourist to have made his approach on the southeast, to the Central or White Mountain group, via Route 4, Lake Winnipiseogee and Conway valley, and thus meet him at the Crawford House, near the

Great Notch. The mountains, which have gradually gathered about us, in our steep ascent here have all closed in. The magnificent pass—the Gateway of the Notch, is a chasm between two perpendicular masses of rock, approaching each other to within 22 feet. Dark overhanging cliffs stand as sentinels over this solemn pass, and it has been a work of toil to cut a pathway through the frowning barrier. This gorge is some three miles long, descending the valley of the Saco, towards "the Willey House." Upon the north, the bold cliffs of Mount Willard rise to the height of 2,000 feet above quiet vales below. The rugged flanks of the devoted Mount Willey, bearing yet the fatal tempest scars of 1826, stops the view on the left, while Mount Webster—dark, and massive, and grand, as was he whose name it bears—fills the landscape on the right.

The White Mountains (specifically so called) extend from the Notch, in a north-easterly direction, some 14 miles, increasing from each end of the line gradually in height towards Mount Washington, in the centre. These respective elevations are, in the order in which they stand, beginning at the Notch—Mount Webster, 4,000 feet above the level of the sea; Jackson, 4,100; Clinton, 4,200: Pleasant, 4,800; Franklin, 4,900; Monroe, 5,300; Washington, 6,500; Clay, 5,400; Adams, 5,700; Jefferson, 5,800; and Madison, 5,400.

Passing westward from the Notch, we reach the valley of the Ammanoosuc, after a distance of four miles, through, dense woods, and enter abruptly into a spacious clearing, from which the whole mountain group bursts upon our wondering sight. Here, upon the "Giant's Grave," an eminence of some 60 feet, the panorama is marvellous. In the centre of the amphitheatre of hills, Mount Washington, barren, and seamed and whitened by the winter tempests of centuries, looks down, upon the right and upon the left, on the hoary heads of Webster and Madison—each, on its side, the outpost of the mountain army.

The Hotels. The Crawford House —a most excellent establishment—bears the name of the earliest hosts of these mountain gorges. The story of the adventures and the endurance of the early settlers here, is extremely interesting. How Captain Eleazar Rosebrook, of Massachusetts, built a house on the site of the Giant's Grave, four miles from the Notch, afterward occupied by Fabyan's Mount Washington Hotel*— how his nearest neighbors were 20 miles away, excepting the Crawford family, 12 miles down in the Notch valley— the present old Crawford House, at the base of the mountains, coming from Conway, on the south-east. How the Rosebrook children were often sent, for family supplies, over the long and dangerous path to the Crawfords', returning, not unfrequently, late at night— how Ethan Allen Crawford was heir to the Rosebrook estate, and how he became known as the "Giant of the Hills" —how he and his family made the first mountain paths,† and were for long

* Destroyed by fire and never rebuilt.
† The first bridle-path was cut by Ethan Crawford, in 1821.

years the only guides over them of the rare visitors, which the brief summers brought—and how they have since seen their home thronged, for weeks together, like a city saloon, with beauty and fashion. The Crawfords are a large, athletic race. Abel, the father, called the "Patriarch of the mountains," would walk five mountain miles, to his son's, before breakfast, at the age of 80. At 70, he made the first ascent ever made on horseback, to the top of Mount Washington. His sons were all over six feet tall; and one of them was six and a half feet, and another, Ethan Allen, was seven feet in height.

Ascent of Mount Washington. The chief exploit of the visitor, at this group of the White Hills, is to ascend Mount Washington; and a toilsome, and even dangerous feat it is to this day, despite the improved facilities of travel. The journey from the Crawford House is nine miles, made on the backs of Canadian ponies, over the old Crawford bridle-paths, though a grand carriage-way is now projected, from the Glen House on the opposite side of the mountain. (See Glen House.) The excursion occupies a long day, with the utmost industry. We made it, on one occasion, in midsummer, with a party of thirty ladies and gentlemen, besides our guides, and it was a gay scene—the getting *en route*, and a singular cavalcade; miles onward as we wound, in Indian file, cautiously along the rugged, narrow path, trusting to our trusty ponies to walk with us upon their backs, over logs, and rocks, and chasms, which we would not have dared to leap ourselves; and surprising was the picture, as we, at length, bivouacked and ate our grateful lunch upon the all-seeing crest of the grand old mountain. At another time, we ascended, in the middle of October, when we could muster no larger group than our friend, ourself, and our guide. For two miles from the summit, the way was blocked with snow; so we left our ponies to take care of themselves, and completed the tramp on foot. The day, though so bitterly cold as to remind us of Webster's salutation upon a like occasion—"Mount Washington, I have come a very long distance, have toiled hard to arrive at your summit, and now you give me but a cold reception"—was happily a brilliant one ; the atmosphere was exceedingly clear : and we had the delight of seeing all the delicious panorama, which has been thus catalogued :—

View from the Summit. "In the west, through the blue haze, are seen, in the distance, the ranges of the Green Mountains ; the remarkable outlines of the summits of Camel's Hump and Mansfield Mountain being easily distinguished when the atmosphere is clear. To the north-west, under your feet, are the clearings and settlement of Jefferson, and the waters of Cherry Pond ; and, further distant, the village of Lancaster, with the waters of Israel's river. The Connecticut is barely visible ; and often its appearance for miles is counterfeited by the fog rising from its surface. To the north and north-east, only a few miles distant, rise up boldly the great north-eastern peaks of the White Mountain range—Jefferson, Adams, and Madison—with their ragged tops of loose dark rocks. A little further to the east are seen the numerous and distant summits of the mountains of Maine. On the south-east, close at hand, are the dark and crowded ridges of the mountains of Jackson ; and beyond, the conical summit of Kearsarge, standing by itself, on the outskirts of the mountains ; and, further over the low country of Maine, Sebago Pond, near Portland. Still further, it is said, the ocean itself has sometimes been distinctly visible.

"The White Mountains are often seen from the sea, even at 30 miles distance from the shore ; and nothing can prevent the sea from being seen from the mountains, but the difficulty of distinguishing its appearance from that of the sky near the horizon.

"Further to the south are the intervals of the Saco, and the settlements of Bartlett and Conway, the sister ponds of Lovell, in Fryburg ; and, still further, the remarkable four-toothed summit of the Chocorua, the peak to the right being much the largest, and sharply pyramidal. Almost exactly south are the shining waters of the beautiful Winnipiseogee, seen with the greatest distinctness in a favorable day. To the south-west, near at hand, are the peaks of the south-western range of the White Mountains ; Monroe, with its two little alpine ponds sleeping under its rocky and pointed summit ; the flat surface of Franklin, and the rounded top of Pleasant, with their ridges and spurs. Beyond these, the Wiley Mountain, with its high, ridged summit ; and, beyond that, several parallel ranges of high-wooded mountains. Further west, and over all, is seen the high, bare summit of Mount Lafayette, in Franconia."

Tuckerman's Ravine is a marvellous place, seen in the ascent of the mountains, by the Davis' Road leading from the Crawford House. It lies upon the right in passing over the high spur directly south-east of Mount Washington. Turning aside, the edge of the precipice is reached, and may be descended by a rugged pathway. It is a long, deep glen, with frowning walls, often quite inaccessible. It is filled, hundreds of feet deep, by the winter snows, through which a brook steals, as summer suns draw near, gradually widening its channel, until it flows through a grand snow cave, which was found, by measurement, to be, one season, 84 feet wide on the *inside*, 40 feet high, and 180 feet long. The snow forming the arch was 20 feet thick! The engineers of the projected carriage road went through this arch one July, in the bed of the stream, to the foot of the cataract, which flows for 1,000 feet, down the wild mountain side.

Oakes' Gulf is another fathomless cavern, seen, far down on the right, in winding around the summit of Mount Monroe. Near the summit of Mount Washington, a few rods northward, is yet another black abyss, which is called the Great Gulf. Its descent here is 2,000 feet, rugged and precipitous.

The Summit House. The building of the rude inn, which now stands upon the summit of Mount Washington where the great hotel is to be when the grand projected road is completed, was a daring and painful enterprise. It is said the bold scheme was suggested by Jos. S. Hall, for many years a guide from the Notch House. "The house was located," says Mr. Willey, "under the lee of the highest rocks on Mount Washington, and was laid out forty feet long, and twenty-two feet wide. The walls were four feet thick, laid in cement, and every stone had to be raised to its place by muscular strength alone.

While these were laying the walls, the materials for finishing and furnishing were being dragged up from the Glen House, a distance of six miles. Lime, boards, nails, shingles, timbers, furniture, crockery, bedding and stoves, all had to be brought up by piecemeal on the men's or horses' backs. No one ever went up without taking something —a chair, or door, or piece of crockery. Four boards (about sixty feet) could be carried up at one on a horse's back, and but one trip could be made daily. Mr. Rosebrook, a *young giant*, carried up at one time a door of the usual length, three feet wide, three and one-half inches thick, ten pounds of pork, and one gallon of molasses.

The walls were raised eight feet high, and to these the roof was fastened by strong iron bolts; while over the whole structure were passed strong cables, fastened to the solid mountain itself. The inside was thrown, primitive fashion, into one room, in which the beds were arranged, berth-like, for the most part on one side of the room, in two tiers, with curtains in front. A table, capable of seating thirty or forty persons, ran lengthwise of the room. At one end of the room a cooking-stove and the other furniture of a kitchen were placed, with a curtain between it and the table. At the other end was a small stove, in which was burned mountain moss. The walls are perfectly rough, outside and in; a little plaster upon the inside, merely fills up the chinks. The Tip Top House was the second structure erected here.

Much as we have necessarily left unseen, on the mountains, we must now descend, and with a hasty peep at some yet unmentioned scenes, in the vicinage of the Notch, pass on, thirty-six miles, to the Franconia range, in the west.

The Silver Cascade is a favorite scene, about half a mile south of the entrance to the Notch. It is one of the most charming waterfalls imaginable, seen at a distance of two miles, bubbling down the mountain side, eight hundred feet above the neighboring valley.

THE FLUME is another cascade yet further down the Notch. It descends two hundred and fifty feet, in two rills, over two precipices, and there are three streams over a land ledge, reuniting in a small rocky basin below.

The Devil's Den is a mysterious cavern, near the top of Mount Willard, opposite the Silver and the Flume cascades.

The Crystal Falls, of eighty feet, and the *Glen Ellis Falls,* of seventy feet, are on the Ellis River, the one on the left and the other on the right of the road from Jackson to the Glen House.

The Ammanoosuc River, rising in this group of the White Mountains, and followed in the journey towards the Franconia Hills, is a stream of wonderful beauty. It falls 6,000 feet from its source on the mountain, to the Connecticut River, and is said to be the wildest and most impetuous river in New Hampshire. It abounds in rapids and cascades.

The Willey House is passed some miles below, at the commencement of the ascent to the Notch. The spot will be for ever of memorable interest, from its tragic story of the fearful avalanches of 1826, when the entire family which then occupied the house—Mr. Willey, his wife, five children and two hired men — were all buried beneath the mighty *debris* of the mountain slides.*

* A circumstantial narrative of this melancholy event may be found in the Rev. Mr.

The ride through the hills and by the waterfalls, 23 cheery miles from the Crawford House to the Profile House, in the Franconia Pass, might detain us pleasantly enough at many points, but we bear our traveller on, at once, to the last chapter of our White Mountain story.

The Franconia Hills, though in popular estimation inferior in interest to the eastern cluster, are really not so; except it be in the wonders of the mountain ascents; and even in this, the panorama, from the summit of Lafayette, is scarcely less extensive or less imposing than the scene from the crown of Mount Washington, while the exquisite little lakes, and the singular natural eccentricities in the Franconia group, have no counterpart in the other. In this, as in other ranges of the White Hills, the mountains are densely wooded at their base, while their rock-ribbed summits are barren, and scarred by time and tempest. The hills approach, at one point, to within half a mile of each other, and form the wild procrustean portal, called the Notch.

Mount Lafayette, or the Great Haystack, is the monarch of the Franconia kingdom, towering up, skyward, to the height of 5,200 feet. Its lofty pyramidal peaks are the chief objects, in all views, for many miles around.

Eagle Cliff is a magnificently bold and rocky promontory, near Mount Lafayette. It casts its dark shadows down many hundred feet into the glen, traversed by the road beneath.

Cannon Mountain, 2,000 feet above the road and 4,000 above the sea, is nearly opposite Lafayette, and forms the western side of the Notch. Away up upon its crown is a group of mighty rocks, which, as seen from the Profile House below, bear an exact resemblance to a mounted cannon. It is upon this mountain, also, that we find that marvellous freak of nature,

W.lley's "Incidents of White Mountain History,"—which we have already quoted and oftener referred to, in verification of our own memories and notes of those mountains. The unfortunate family were near relatives of Mr. Willey, and he lived in the vicinity at the time of the disaster.

The Profile Rock, or The Old Man of the Mountain.—This wonderful eccentricity, so admirably counterfeiting a human face, is 80 feet long, from the chin to the top of the forehead, and is 1,200 feet above the level of the road, being yet far below the summit of the mountain. This strange apparition is formed of three distinct masses of rock, one making the forehead, another the nose and upper lip, and a third the chin. The rocks are brought into the proper relation to form the Profile, at one point only, upon the road, through the Notch, a quarter of a mile south of the Profile House. The face is boldly and clearly relieved against the sky, and, except in a little sentiment of weakness about the mouth, has the air of a stern, strong character, well able to bear, as he has done unflinchingly, for centuries, the scorching suns of summer and the tempest blasts of winter. Passing down the road a little way, the "Old Man" is transformed into a "toothless old woman in a mob cap;" and, soon after, melts into thin air, and is seen no more. Hawthorne has found in this scene the theme of one of the pleasantest of his "Twice Told Tales," that called the "Old Stone Face."

The Profile Lake is a beautiful little pond, lying at the base of the mountain, and immediately under the ever-watchful eye of the stern "Old Man." This lakelet is sometimes called the "Old Man's Wash-bowl." It is a quarter of a mile long and about half as wide.

Echo Lake, one of the greatest charms of this part of the White Mt. region, is a diminutive but very deep and beautiful pond, north of the Cannon Mountain. It is entirely enclosed by high mountains. From the centre of this fairy water, a voice, in ordinary tone, will be echoed distinctly several times, and the report of a gun breaks upon the rocks like the roar of artillery. The Indian superstition was, that these echoes were the voice of the Great Spirit, speaking in gentleness or in anger.

The Basin, another remarkable

The White Mountains.

Profile Lake, White Mountains.

scene of this neighborhood, is five miles south of the Notch. It lies near the road side, where the Pemigewasset has worn deep and curious cavities in the solid rocks. The Basin is 45 feet in diameter, and 28 feet from the edge to the bottom of the water. It is nearly circular, and has been gradually made by the whirling of rocks round and round in the strong current. The water, as it comes from the Basin, falls into most charming cascades. At the outlet, the lower edge of the rocks has been worn into a very remarkable likeness of the human leg and foot.

The Flume, the last and most famous, perhaps, of all the Franconia wonders, is quickly and easily reached from the Flume House. Leaving the road, just below the Basin, we turn to the left among the hills, and after a tramp of a mile, reach a bare granite ledge a hundred feet high and about thirty feet wide, over which a small stream makes its varied way. Near the top of this ledge we approach the ravine known as the Flume. The rocky walls here are fifty feet in height, and not

more than twenty feet apart. Through this grand fissure comes the little brook which we have just seen. Except in seasons of freshets, the bed of the stream is narrow enough to give the visitor dry passage up the curious glen, which extends several hundred feet, the walls approaching, near the upper extremity, to within ten or eleven feet of each other.

About midway, a tremendous boulder, several tons in weight, hangs suspended between the cliffs, where it has been caught in its descent from the mountain above. A dangerous bridge for a timid step has been sprung across the ravine, near the top, by the falling of a forest tree.

The Pool, a supplemental or tail piece to the great picture of the Flume, is a deep natural well in the solid rock. The diameter of the Pool is about sixty feet; the depth to the surface of the water is 150 feet, and the water itself extends 40 feet yet below. Some years ago, a poor fellow was unlucky enough to fall into this plutonian *cul de sac*, but he clung to a crag just above the water until ropes were lowered, and he was, wonderful to relate, fished up, though bruised and not a little scared, yet alive!

———

We have now peeped hastily at the leading points of interest in the grand Granite Hills; but the enterprising tourist, of an inquiring turn of mind, may, very easily, discover for himself a thousand other marvels and delights; or, if he cares not to explore new scenes for himself, he may bend his way northward, via Littleton, and between Colebrook and Errol, penetrate the well-known, but as yet very little visited slate-stone gorge, called the Dixville Notch.

VERMONT.

THE thousand points of interest in the Green Hills of Vermont have not yet received their due meed of favor from tourists, but their claims to especial homage are now being fully admitted. The mountain chain extends from near New Haven, in Connecticut, northward through Massachusetts and Vermont, into Canada; though, properly speaking, it lies in Vermont alone, where are the chief summits of Mansfield, Camel's Hump, Connell's Peak, Shrewsbury Mountain, South Peak, Killington Peak, Ascutney (on the Connecticut), and others. After the White Mountains of New Hampshire, the Green Hills rank with the noblest groups west of the Rocky Mountains—with the Blue Ridge in North Carolina, Georgia, and Virginia, the Alleghanies in Pennsylvania, and the Catskill and the Adirondacks, New York.

The Vermont Central Railway from Burlington, on Lake Champlain, eastward via Montpelier, the capital of Vermont, to the shores of the Connecticut River, traverses the valley of the Winooski, by the banks of the Winooski River, and gives easy access to Mount Camel's Hump and Mount Mansfield. **The Valley and River of Winooski.**—The Winooski traverses almost the entire breadth of northern Vermont. Rising in Caledonia, its course is generally westward to Lake Champlain, 40 miles from which it passes through Montpelier, the capital of Vermont. The Vermont Central Railway from Burlington to the Connecticut runs through the valley, and very closely follows the banks of the river as far eastward as Montpelier. Some of its valley passages are scenes of great pastoral beauty, strongly contrasted with high mountain surroundings, the singularly-formed peak of Camel's Hump continually re-occurring, sometimes barely peeping over intervening ranges, and again—as near the middle of the valley stretch—coming into full display. In places, the Winooski is a wild turbulent water, dashing over stern precipices and through rugged defiles. It is found in this rough mood just above the village of Winooski, a few miles from Burlington, where the waters rush in rapid and cascade through a ravine a hundred feet. This picture is well seen from the railway. Passing on into the open valley lands, which succeed, Mount Camel's Hump comes finely into view, as the central and crowning point of one of the sweetest pictures of all this region. Continuing always through scenes of great picturesque interest, the tourist comes to the village of Richmond, 13 miles from Burlington, and three miles, yet beyond, to Jonesville, a little scattered village of fine farms, lying upon both sides of the river. The inn here was a famous halfway-house in the journey between Burlington and Montpelier, before the railway was built, and it is still a pic-nic

Valley of the Winooski.

Falls on the Winooski River, Vermont.

and excursion resort for all the region. It is an excellent and very inexpensive place for a little quiet tarry. Mount Camel's Hump is accessible from this vicinity, and a stage runs from the hotel, some nine or ten miles, to Underhill Centre, at the foot of Mount Mansfield. On the south side of the Winooski, at Jonesville, the Huntington River comes in, after an exceedingly wild journey for the last two or three miles of its course, through fine rocky ravines, which it traverses with many bold foaming leaps. About three miles yet east of Jonesville, near Bolton, there is the most striking picture in all the Winooski gallery. It is beautifully seen from the cars on the left, but yet very inadequately. From the bottom of the glen through which the maddened waters here make their way, the huge cliffs on either hand, the torrent foaming at their feet, and the gentle bit of verdant interval, with the tall crest of Camel's Hump, seen beyond, within the frame of the opposing precipices, make altogether a scene well worth the looking for. The pictorial interest of this valley transit is admirably sustained at all points, as far as we yet follow it onward, through Waterbury and Middlesex, to Montpelier, where some of the best scenes the traveller will find, when he looks abroad from the hill-tops there, have been kept to the last.

Mount Camel's Hump, the most salient feature in the Winooski landscape, is the highest of all the Green Mountain peaks, having an elevation of 4,188 feet. It may be ascended, without much difficulty, from any side, though the usual point of access is at Duxbury, from whence carriages can pass to within three miles of the summit. The mountain is crowned by jagged, barren rocks, and the imposing scene which the lofty heights overlook is in no way obstructed by the forest veil, which often disappoints the hopeful climber of forbidding mountain tops.

Mansfield Mountain, the second in dignity of the Green Hills, is very accessible from the village of Underhill Centre on the north, or yet more easily from Stow on the south, both of which points may be reached from the Vermont Central road—Underhill, from Jonesville station, and Stow from Waterbury. The views of the mountain itself, its cliffs and peaks, are very grand from many points in the path upwards, and the panorama unfolded upon the summit is, if possible, finer than that from the Camel's Hump. Lake Champlain and the Adirondack peaks lie to the westward, while the White Mountains of New Hampshire make yet new pictures on the east; and, again, the many crests of the Green Hills, with their intervening vales and lakes and villages, stretch out towards the south.

Montpelier on the east, and **Burlington** on the west, are the two extremities of the Winooski section of the Green Mountain scenery. Burlington is upon the Lake Champlain shore, and is the largest town in Vermont. We have already mentioned it in our tour from New York to Canada. Montpelier is the State capital. It has a population of between two and three thousand; is a very pretty town, and with the charm of most beautiful natural surroundings, and the comforts of an excellent hotel, it is perfectly eligible as a summer residence. The Winooski river passes here. A little branch railway, of a mile or so, connects Montpelier with the Central road. The State House, which was destroyed by fire January 6th, 1857, was an imposing granite edifice, in the form of a cross. It was 150 feet in length, and 100 deep. It had a fine portico of six columns, each six feet in diameter at the base, and 36 feet high. The apex of the dome which crowned the structure was

Burlington, Vermont.

100 feet from the ground. Among the objects of interest in this edifice were two pieces of cannon taken from the Hessians at the battle of Bennington.

Rutland to Burlington.

Rutland is near the western borders of Vermont, south of the centre of the State, and nearly east of Whitehall, at the lower extremity of Lake Champlain. It is a centre of railway lines for all points of the compass. The Troy, Whitehall, and Castleton R. R., 95 miles, unites Rutland with Troy and Albany, via Whitehall and Saratoga Springs. It is also connected, again, with Troy and Albany, by the Albany, Vermont, and Canada line (formerly Albany Northern) to Eagles Bridge, and thence by the Rutland and Washington road, 95 miles, as via Whitehall and Saratoga; and yet again, via North Bennington, by the Troy and Boston and Western Vermont roads, 84 miles. The famous Hoosick tunnel is near the line of this route. Northward, it is connected with Burlington, and all the routes which intersect at that point, by the Rutland and Burlington R. R., 50 miles, and eastward with Bellows' Falls, on the Connecticut, by another division of the same line. Rutland is a pleasant town, with a population of about 4,000, situated in the midst of some of the finest of the Vermont hill and valley scenery, at the foot of the western slope of the mountains. Otter Creek, a most picturesque stream in all its course, passes by the village, and Killington Peak is admirably seen as the leading feature in the landscape around.

The Clarendon Springs, a favorite place of resort, is a few miles south of Rutland, on the line of the Western Vermont road. The medical virtues of these waters, the varied and beautiful scenery, the pleasant drives around, and the excellent hotel accommodations, make this watering-place a very desirable summer halt.

The Otter Creek Falls, at Vergennes, are upon the Otter Creek, about seven miles from Lake Champlain.

The brook is 500 feet in width, divided by a fine island, on either side of which the Fall leaps bravely some 30 or 40 feet. There are many other beautiful cascades in the Otter Creek. Some at Middlebury, above Rutland; and a few miles below Middlebury, still others of yet greater interest.

The Elgin Spring is in the neighborhood of the Otter Creek cascades.

Killington Peak, rising grandly on the east of Rutland, is the third in rank of the mountains of Vermont. A visit to this peak makes a pleasant excursion from the neighborhood. To the foot of the mountain the distance is seven miles, and two miles more to the summit. On the north side is a perpendicular ledge of 200 feet, called Capitol Rock. Mount Ira, too, is hereabouts, and beyond Killington Peak, as seen from Rutland, and northward is Mount Pico and Castleton Ridge, shutting out the view of Lake Champlain.

Lake Dunmore is a delicious water, 30 miles above Rutland. It is on the railway to Burlington, a few miles (by stage) from Middlebury. Dunmore is a wonderfully picturesque lake, surrounded at most points by bold hills, seen here in verdant slopes, and there in rocky bluff and cliff. The lake is about five miles in length and three in breadth. A good summer hotel is upon its banks.

Lake Castleton, in this neighborhood, is also a most interesting scene.

Eastward from Rutland, the route lies amidst the beauties of the Queechy Valley, replete with delightful pictures of running and falling waters, and of grassy meadows and wooded hills.

Middlebury is upon the railway to Burlington, 30 miles from Rutland. It is upon the Otter Creek, near some fine falls on that stream, and is also a few miles only from Lake Dunmore. It has a population (the township) of some 4,000, and, like nearly all the villages in Vermont, is a very beautiful place, surrounded at all points by most attractive mountain scenery. It is distinguished as one of the first manufacturing towns in the State, and also as the seat of

Middlebury College. This institution was founded in 1800. Its chief edifice is 100 feet long and four stories high, built of stone.

Brandon, on the route of the Rutland railroad, is a flourishing town, finely watered by Otter Creek, Mill River, and Spring Pond, on which streams are good mill-seats. Minerals of fine quality are found in this town. There are here two curious caverns formed of limestone, the largest containing two apartments, each from 16 to 20 feet square. It is entered by descending from the surface about 20 feet.

Bennington is at the meeting of the Troy and Boston and the Western Vermont railways, in the extreme southwest corner of the State. It is famous as the scene of the battle of Bennington (August 16, 1777), in which a detachment of the enemy's forces, under Gen. Burgoyne, was terribly beaten by the Green Mountain Boys, led by the intrepid Major Stark. It was upon the occasion of this memorable engagement that Stark is reported to have made the famous address to his troops: "See there, men! there are the red coats! Before to-night they are ours, or Molly Stark will be a widow!" Two pieces of artillery, taken in the battle of Bennington, were, until recently, preserved in the Capitol at Montpelier. The manufactories of the United States Pottery Company at Bennington are well deserving of a visit. Fine porcelain and Parian ware are made here, the vicinage yielding the necessary materials in abundant and excellent supply. The landscape about Bennington is not of especial attraction.

Willoughby Lake is a popular resort in Orleans County, Vermont, lying upon the Canada line. This lake is a beautiful water, nearly five miles long. It lies upon the great railway route from Boston, via St. Johnsbury, to Canada, leaving the Connecticut Valley route at Wells River.

Lake Memphremagog is 30 miles long, and from one to four in breadth. About eight miles only of its waters are in Vermont, the rest lying in Canada. It unites its waters, by Magog outlet, with those of the St. Francis river, in Canada.

For Mount Ascutney, Windsor, Bellows' Falls, Brattleboro', and other scenes and places in Vermont, on and near the Connecticut River, see route through that region.

NEW YORK.

New York is very appositely called the Empire State; being the first in the confederacy in population, first in wealth, and in commercial importance; exceeded by none in soil and climate, unsurpassed in the variety and beauty of her natural scenery, and in her historical associations.

The earliest settlements here, were made by the Dutch, at Fort Orange, now called Albany, and at New Amsterdam, now New York City. This was in 1614, seven years after the voyage of Hendrick Hudson up the waters of that river which now bears his name.

In 1664, the colony fell into the possession of the English—was recaptured by the Dutch in 1673, and finally came again under British rule in 1674—and so continued until the period of the Revolution. Many stirring events transpired within this territory during the wars between France and England, in 1690, 1702, and 1744, and through all the years of the War of Independence. Of these events the traveller will find some chronicle as we reach the various locations where they transpired, hereafter, in the course of our proposed travels:—

Every variety of surface and every character of physical aspect is found within the great area of New York; vast fertile plains and grand mountain ranges; meadows of richest verdure, and wild forest tracts; lakes innumerable and of infinite variety in size and beauty; waterfalls unequalled in the world in extent and grandeur; and rivers matchless in picturesque charms. We need not now catalogue these wonders, as our rambles will afford us, by and by, abundant opportunity to see them all in turn and time—the peaks and gorges of the Adirondacks and the Catskills—the floods of Niagara and the ravines of Trenton; the pure waters of Lake George, the mountain shores of Champlain, the deer-filled wildernesses and the Highland passes of the Hudson, and all the intricate reticulation of cities, towns, villages, villas, and watering-places.

RAILWAYS IN NEW YORK.

See index for description of the routes.

The New York and Erie, 459 miles through the State, from the city of New York to Dunkirk, or 422 to Buffalo (Branch) on Lake Erie. A route to the far West.

The Hudson River Railway, 146 miles to Albany, or 152 to Troy, along the banks of the Hudson River, from New York city.

The Harlem Railway, 154 miles from New York to Albany.

New York Central, from Albany to Buffalo, 398 miles, or to Niagara Falls, 327 miles — unites eastward with the Western Railway from Boston, and with the Hudson River and Harlem roads from New York and at the western extremity, with routes for the Mississippi regions.

Rensselaer and Saratoga and Saratoga and Whitehall. From Troy to Saratoga Springs, 32 miles; to Whitehall, 72 miles.

Troy and Boston, and Albany and Rutland railways.

Montreal and New York, and Plattsburg and Montreal railways, 62 miles from Plattsburg, on Lake Champlain to Montreal, Canada.

Northern (Ogdensburg) Railway, across the northern part of the State, 118 miles, from Rouse's Point, on Lake Champlain, to Ogdensburg on the St. Lawrence.

Black River and Utica, 109 miles from Utica, on the New York Central Road to Clayton, St. Lawrence, near Lake Ontario.

Watertown and Rome; 97 miles from Rome on the New York Central, to Cape Vincent, on Lake Ontario.

Potsdam and Watertown, from Watertown junction (Watertown and Potsdam Road) to Potsdam, on the Northern (Ogdensburg) Railway.

Newburgh Branch of New York and Erie Railway; from Newburgh on the Hudson, to Chester, New York and Erie Railway.

Oswego and Syracuse; 35 miles from Syracuse, New York Central Road, to Oswego, Lake Ontario.

Syracuse and Southern; 80 miles from Binghampton (Erie Railway) to Syracuse (New York Central Railway).

Oswego (New York and Erie Railway), 35 miles to Ithaca, on Cayuga Lake.

Elmira, Canandaigua and Niagara Falls; 168 miles from Elmira (Erie Railway), to Suspension Bridge, Niagara.

Buffalo, Corning and New York; from Corning (Erie Railway), 100 miles to Batavia, or 94 miles to Rochester (New York Central Road).

Williamsport and Elmira; 78 miles from Elmira (Erie Railway)—south to Williamsport, Pa.

Corning and Blossburg and Tioga; 41 miles from Corning (Erie Railway) to Blossburg, Pa.

Lake Shore Railway, from Buffalo, via Dunkirk, by the shore of Lake Erie, to Cleveland, Ohio, and westward.

Hudson and Boston; from Hudson, on the Hudson River, eastward, to West Stockbridge, 34 miles.

Western Railway; from Albany, 200 miles, to Boston.

New York and New Haven; 75 miles from New York to New Haven, Ct., thence to Boston, etc.

Long Island Railway; 95 miles from New York (Brooklyn Ferry), through

CITI
NEW-
BROOK

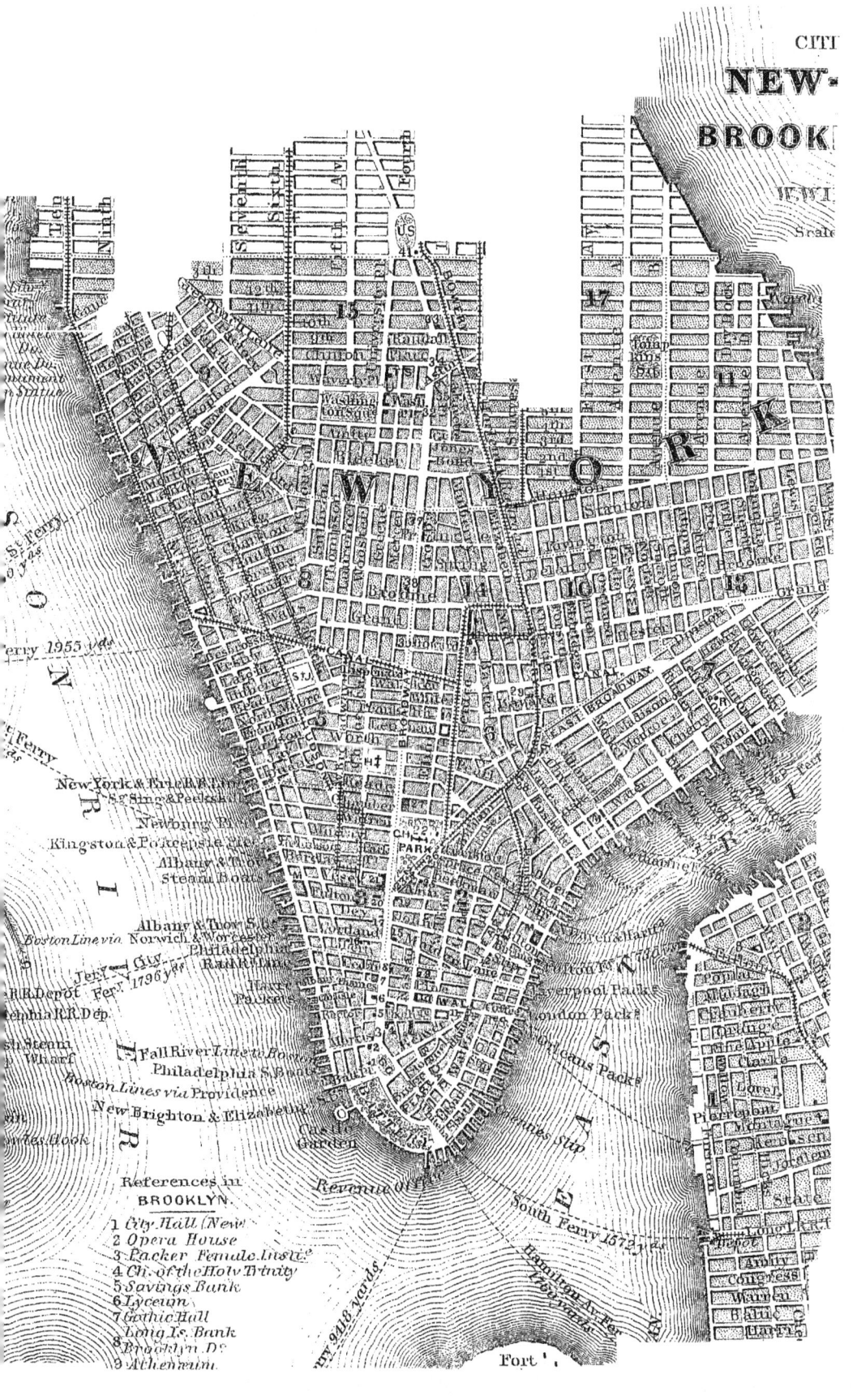

NEW YORK

BROOKLYN

References in
BROOKLYN.
1 City Hall (New)
2 Opera House
3 Packer Female Inst.
4 Ch. of the Holy Trinity
5 Savings Bank
6 Lyceum
7 Gothic Hall
8 Long Is. Bank
9 Athenæum

Castle Garden

Fort

the entire length of Long Island, to Greenport.

THE CITY OF NEW YORK.

Supposing our traveller to be at home in New York, or a stranger already comfortably lodged in his hotel, we will (leaving it to a later moment to see whether or not he has got into the right place), gossip, for a brief while, touching the past of the scenes, of which we propose to show him the present.

The rapid growth of this great city— so little time gone by a wild, forest settlement, and now magnificent in its million people—is evidence enough of the mind and will of the race, which is now every where making the once wilderness of the west to blossom as the rose. Though settled by the Dutch as early as 1612, the metropolitan character of New York scarcely dates back to the beginning of the present century : for it is within the past 50 years, or less, that all its present municipal glories and fame have grown up. Not so long, indeed, for the city which now covers nearly the whole Island of Manhattan, and is running over every day into other cities, villages, and suburbs, wherever it can find vent, was, within a shorter period, composed in the small triangular area, of which the Battery is the apex, and Canal street the base. The City Hall was then built with less care, on the upper or north side, because that, at the time, overlooked, and was seen only from lanes and fields. To go *above* or even to Canal street, (then, literally, a canal), was a rural excursion ; while, to go *below* it, at this day, is to exceed the general down-town travel, on any but business errands. Of the palatial private houses, the public structures, the magnificent churches, the parks, and even the streets, in all the middle and upper parts of the City, no mention would have been made in an edition of this work twenty-five years ago ; which leads us to ask, what story it may be necessary to tell, in the revised edition of a quarter of a

century hence ! But, thinking no longer of past or future, let us come at once to the present, and see New York as it is.

Panorama of the City. The visitor will do well to accompany us to the lofty outlook from the top of the Tower of Trinity Church, in the lower part of Broadway. Here he can pick up some general idea of the topography and extent of the City. He may go there at any time when the building is not in use for sacred service, paying the porter a fee for his guidance. At the landing, on a level with the ceiling of the church, a fine view is had of the beautiful interior. At the head of another flight of stairs, the belfry, with its pleasant chimes, is reached. Here, too, is a balcony affording a fine view of the City ; but it is still higher up that the scene is spread forth in all its glory—a boundless array of charms, in city, and town, and village, river, and bay, and island, all teeming with bright and busy life and action.

With this superb picture, or rather galleries of pictures, before him, the observer gets a better idea than he may, perhaps, have had before of one of the natural advantages which has made New York the great metropolis of this wide country ; its noble position at the meeting of great waters, leading inland, and its unrivalled harbors upon the sea.

Yonder stretches the beautiful bay— one of the safest and easiest of access in the world—eight miles out to those great portals famous as the "Narrows," which open its way to the ocean. The circumference of the harbor is 25 miles, within which, the combined navies of the world might lie in comfort and security. The scenery here is of infinite attraction, in all the protean shapes and effects of mingled land and water. Great ships and little crafts innumerable seem to jostle each other, and cities, and villages, and villas crowd the shore, from the water's edge to the bold hill-tops. The outer harbor, or the bay proper, extends from the Narrows to Sandy Hook Light, 18 miles from the city. Within the harbor are the picturesque fortifications on Governor's, on Bedlow's, and on Ellis's Islands. Fort Columbus occupies the centre of Governor's Island, and at its north-east point is Castle William, a round tower some 600 feet in circumference, and 60 feet high, with three tiers of guns; while at the north-west is a battery, commanding the entrance to Buttermilk Channel, by which the island is separated from the City of Brooklyn. The defences on the Long Island shore, at the Narrows, are Forts Hamilton, and Lafayette—formerly Fort Diamond. This neighborhood is a popular summer resort and residence of the people of New York. Opposite these fortifications, on the Staten Island shore, separated here by the passage of the Narrows, about two-thirds of a mile in width, are Forts Tompkins and Richmond.

Staten Island, a favorite suburban home of New York, and to which the Bay is indebted for so much of its beauty, is about six miles below the city, with which it has frequent daily connection. The island is 14 miles long, and from four to eight wide. It constitutes the county of Richmond, and forms the southern extremity of the State of New York. It is separated from New Jersey, on the west, by Staten Island Sound. Richmond Hill, at the north end of the island, commands all the grand scenes which might be expected in this vicinage, at an elevation of 307 feet above the sea. Elegant residences cluster about these heights, and, from the summit, a marine telegraph overlooks them and the sea. Upon a bluff, on the east side of the island, is Prince's Bay Light House.

Hotels. With this glimpse over the city and harbor, let us now see how our traveller is lodged, for, with the finest and best-appointed hotels in the world at his service, it would be a pity if he were domiciled other than most comfortably.

The Astor House (opened in 1836), we should, perhaps, mention first in our list, as the oldest, and one of the very best of those of the highest rank. It

is a massive granite structure, of simple but imposing design. Its façade extends from street to street, 201 feet, on Broadway, opposite the Park, with its City Hall, and other public buildings, and its great Croton Fountain. It is the only leading hotel left in the lower part of the city, though all the region there is still thronged with most excellent establishments, in favor with citizens and strangers, who consult their business convenience only.

The St. Nicholas, one of the most splendid of all the hotels constructed in New York, is in Broadway, between Broome and Spring streets. Its entire front, according to the original plan, is constructed of pure white marble; but it now includes the new brown-stone building adjoining it, extending up to Spring street. It is six stories high, with 200 feet front on Broadway, and 200 feet in depth.

The Metropolitan Hotel is situated on the corner of Broadway and Prince street, and occupies nearly an entire block. In the rear of this hotel is Niblo's Theatre.

The Fifth Avenue Hotel is a palatial marble edifice, of the most imposing extent and of the highest fashion. It covers the ground between 23d and 24th streets on Fifth avenue, at its intersection with Broadway—opposite Madison square. Rapidly as the large hotels have been creeping "up town" of late, the "Fifth Avenue" outstrips all other strides, and stands now higher up on the island than any of its competitors. This superb house was opened to the public in the fall of 1859.

The Prescott House is a convenient and elegant building, 50 feet on Broadway, and 125 feet on Spring street. It is six stories high. The first story is supported by cast-iron columns, and the upper portion is fashioned of the finest brick.

The Lafarge House is a large and sumptuous hotel of the highest class. It has an imposing façade of white marble upon Broadway, between Bond and Amity streets. It was entirely destroyed by fire at the moment of its completion, but was immediately re-built, and opened, for the first time, in the spring of 1856. The "Winter Garden" Theatre is in the rear, with entrance on Broadway, through the hotel edifice.

The New York Hotel is one of the largest and most fashionable in the city. It is " up town," on Broadway, extending from Washington to Waverley Place. Near by, is the New York University, and the Washington Park. Directly opposite is the Church of the Messiah and Hope Chapel.

The St. Dennis, corner of Broadway and Eleventh street, of unique architecture, is one of the best appointed, and most fashionable houses in the city.

Union Place Hotel is an elegant establishment upon the corner of Broadway and Union Square. In front, northward, is the bronze equestrian statue of Washington, erected July 4th, 1856, the Union Park and fountain.

The Everett House, erected within the past few years, is an imposing, ornamented brick edifice fronting on Fourth avenue and the north line of Union Square. It is directly across, opposite the Union Place Hotel. It has both *table d'hôte* and restaurant.

The Clarendon is yet higher up, on Fourth avenue, corner of Eighteenth street. It is a *recherché* house of high fashion, in especial favor with English travellers.

The Brevoort House, Fifth avenue and Clinton Place, and the St. Germain, Fifth avenue, Broadway, and Twenty-second street, are new up-town hotels of the highest rank. They are, like the Everett House, conducted on the European plan, with both *table d'hôte* and café, rooms and board, or either alone. The Julian, in Washington Place, near Broadway, is a fashionable hotel and boarding-house.

There are, besides, very many most excellent hotels, and hundreds of the second and third class—but we have mentioned enough for all the uses of the travelier.

City Parks and Squares. The

Central Park is a new public domain of such grand extent that it will, before many years, rival the most famous places of the kind in the world. It is situated in the upper part of the city, between the Fifth and the Eighth Avenues, East and West, and 59th and 110th streets, South and North—a noble area of 843 acres, extending 2½ miles in length by ½ a mile in breadth. Millions were expended in the purchase of the ground, and millions more will soon have been generously laid out in embellishing it. The work of improvement—with the help of an army of 3,000 laborers—goes on so fast and so magically that the avenues, and drives, and walks, hills and dales, and lawns and lakes, already make it a very popular resort. In the winter time its frozen ponds are covered with myriads of rollicking skaters, many of whom are of the *beau sex*. Access—(direct) by the city railways—the 6th, 8th, or the 3d avenue roads; fare 5 cents.

The Battery, which contains about 11 acres, is situated at the extreme south end of the City, at the commencement of Broadway, and is planted with trees and laid out in gravel walks. From this place is a delightful view of the harbor and its islands, of the numerous vessels arriving and departing, of the adjacent shores of New Jersey, and of Staten and Long Islands. Castle Garden, on the Battery, was at one time a popular public hall. Here Jenny Lind first sang in America. Here, too, the fairs of the American Institute were once held. It is given over now to the Emigrant Office for a receptacle of the debarking foreign populations.

The Bowling Green, situated near the Battery, and at the commencement of Broadway, is of an oval form, and surrounded by an iron railing. Within its enclosure is a fountain, the water of which falls in pleasant whispers, to the dusty streets, of the freshness and beauty of forest-wilds.

The Park is a triangular enclosure in the lower part of the city; it has an area of 11 acres, containing the City Hall and other buildings.

Union Square, New York.

St. John's Park. Small but beautiful grounds in Hudson street, belonging to the vestry of Trinity Church. St. John's Church, a Chapel of Trinity, is on the east side of the square.

Washington Square, is a pleasant up-town park, a little west of Broadway, with the elegant private residences of Waverley Place and Fourth street on the north and south sides, and upon the east the grand marble edifice of the New York University, and Dr. Hutton's beautiful Gothic church. A superb fountain occupies the centre of these grounds.

Union Park, a most charming bit of wood and lawn, is in Union Square, at the bend in the upper part of Broadway, extending from Fourteenth to Seventeenth streets. On the south-east corner of Union Square is the Union Place Hotel, and the fine bronze equestrian statue of Washington by Henry K. Brown. On the upper side is the Everett House, and, near by, the Clarendon Hotel. Upon the west is Dr. Cheever's "Church of the Puritans."

Gramercy Park is a little to the north-east of Union Square, a charming ground, belonging to the owners of the elegant private homes around it.

Stuyvesant Park is divided in the centre by the passage of the Second avenue. It extends from Fifteenth to Seventeenth streets. The Saint George's Church (Rev. Dr. Tyng), is upon the west side of this park.

Tompkins Square, one of the largest parks of New York, is between Avenues A and B, and Seventh and Tenth streets.

Madison Square is up town, just above the intersection of Broadway and Fifth avenue at Twenty-third street.

Hamilton Square, newer ground still above.

PUBLIC BUILDINGS—MUNICIPAL.

The City Hall is an imposing edifice; the south front is built of marble, and the rear, or north side, of Nyack freestone. It was constructed between the years 1803 and 1810. It occupies the centre of the Park, in the lower part of the city, and is surrounded by other city offices. It is at present proposed to enlarge it very greatly. In this building are twenty-eight offices, and other public apartments, the principal of which is the Governor's room, appropriated to the use of that functionary on his visiting the city, and occasionally to that of other distinguished individuals. The walls of this room are embellished with a fine collection of portraits of men celebrated in the civil, military, or naval history of the country. In the Common Council room is the identical chair occupied by Washington when President of the first American Congress, which assembled in this city.

The *Merchants' Exchange*, in Wall street, is built in the most durable manner, of Quincy granite, and is fire-proof, no wood having been used in its contruction, except for the doors and window frames. It is erected on the site occupied by the Exchange building destroyed by the great fire of 1835. The present one, however, covers the entire block, and is 200 feet long by 171 to 144 wide, and 124 to the top of the dome. Its entire cost, including the ground, was over $1,800,000.

The *Custom House* is situated on the corner of Wall and Nassau streets. It is built of white marble, in the Doric order, similar in model to the Parthenon at Athens. It is 200 feet long, 90 wide, and 80 high. The great hall for the transaction of business is a circular room, 60 feet in diameter, surmounted by a dome, supported by 16 Corinthian columns, 30 feet high, and having a skylight, through which the hall is lighted.

The *Post Office* is in Nassau street, between Cedar and Liberty streets. The building is in no way remarkable for any architectural beauty, but merely as being one of the remnants of the past, having been formerly used as a church by one of the old Dutch congregations.

The *Hall of Justice* or " Tombs," is located in Centre street, between Leonard and Franklin streets. It is a substantial-looking building, in the Egyptian style of architecture, 253 feet long, and 200 wide, constructed of a light-colored granite.

Literary Institutions and Libraries.

Columbia College has been recently removed from the foot of Park Place, near Broadway, far up town, having resigned the old grounds which it has occupied for so many years. The extension of Park Place has already destroyed the ancient green lawns, and its venerable buildings. Columbia College was chartered by George II. in 1754, under the title of King's College. Students, 150. Library, 16,000 vols.

The New York University occupies a grand Gothic edifice of white marble, upon the east side of Washington Park. This structure is a fine example of pointed architecture, not unlike that of King's College, Cambridge, England. The chapel—in the central building—is, with its noble window, 50 feet high, and 24 feet wide, one of the most beautiful rooms in the country. The whole edifice is 180 feet long ; founded in 1831.

The Free Academy, Lexington avenue and 23d street, up town, reached by Harlem cars or Fourth avenue stages. This is a public collegiate academy of the highest rank. Its students are chosen from the pupils of the public schools only. The building is a fine structure, in the style of the town halls of the Netherlands. It will accommodate 1,000 pupils.

The Cooper Union occupies a magnificent brown stone edifice opposite the Bible House on Astor Place, at the point where the union of the Third and Fourth avenues forms the Bowery. This establishment is familiarly known as the Cooper Institute. It was founded by the generous munificence of Peter Cooper, Esq., an eminent merchant of New York. The building erected for its uses cost about 600,000 dollars. It is devoted to the free education of the *people* in the practical arts and sciences. It was publicly opened in Nov. 1859, with over 2,000 students. It contains a noble free reading room. One of its departments is a School of Design for women.

The General Theological Seminary of the Protestant Episcopal Church, is charmingly situated on West 20th st., between Ninth and Tenth avenues.

St. Francis Xavier, 39 West 15th street.—Union Theological Seminary, 9 University Place, just above the New York University.

The New York Historical Society occupies a dainty edifice on Second av., cor. of 11th st. Its library is large and

The Free Academy, New York City.

valuable ; besides which it possesses a fine collection of works of art.

American Geographical Society has rooms in the New York University.

Lyceum of Natural History, in the building of the New York University Medical School, 14th street, near Fourth avenue. Incorporated in 1808, for scientific advancement. The Society possesses a large library, and a fine cabinet of mineralogical specimens.

New York Law Institute. City Hall.

The Astor Library is a public collection of high order, founded by the munificence of the late John Jacob Astor. It numbers at present about 110,000 volumes. The building, on Lafayette Place, is one of the chief architectural attractions of the city.

The Mercantile Library, Clinton Hall (late Astor Place Opera House), Eighth street, near Broadway. This old and popular Institution has at present some 54,000 volumes, in every department of letters. Its members number between 4,000 and 5,000. The winter courses of lectures before the Mercantile Library Association are among the greatest pleasures of the season.

New York Society Library now occupies a new and beautiful building in University Place, near 12th street. It possesses about 36,000 books.

Apprentices' Library (14,000 vols.), is in the Mechanics' Hall, Broadway, near Grand street.

The American Institute is at present in apartments on the first floor of the Cooper Institute in Astor Place. The Annual Exhibitions of mechanic art and industry of this Society, make a feature in the autumn pleasures of the metropolis.

The Mechanics' Institute has a library of about 3,000 volumes ; 20 Fourth avenue.

ART SOCIETIES AND GALLERIES.

The National Academy of Design—the chief Art institution of America—was founded in 1826, since which time it has steadily advanced in influence and usefulness. It numbers among its academicians and associates nearly all of the eminent artists of the city and vicinity. It supports free schools for the study of the antique and the living model; possesses an extensive and valuable Art library; makes Annual Exhibitions of original works by American and foreign painters and sculptors, &c. The Exhibitions of the National Academy are the great event of the spring season in New York. The Academy occupies at present temporary galleries in Tenth street near Broadway.

Studios. In Tenth st. near the Sixth av. there is a spacious quadrangular edifice, called the Studio Building, occupied entirely by artists. A fine gallery, for the uses of the fraternity fills the court. Dodworth's, just above Grace Church on Broadway, is another famous lair of the knights of the easel, and so too is the University in Washington Park. The artist brotherhood of New York is large, and potent in character, both socially and professionally. The stranger or the citizen may while away pleasant days and weeks in exploring their lofty abodes.

The Dusseldorf Gallery is an admirable exhibition of the works of German painters, chiefly of the Dusseldorf school — 548 Broadway, in the front building of Dr. Chapin's church.

The Free Fine Art Gallery of Messrs. Williams, Stevens, Williams and Co., 353 Broadway, is always rich in objects of rare interest, pictures, engravings, and other works.

The New Bible House is one of the largest structures in the city. It covers the entire area between Third and Fourth avenues on the west and east, and 8th and 9th street on the south and north. The printing rooms and other offices of the American Bible Society are here.

The New York Hospital (founded in 1771) stands back on a lawn, upon Broadway, opposite Pearl street.

Medical Schools. New York University Medical Department, 107 East Fourteenth street. *College of Physicians and Surgeons*, East Twenty-third street and Fourth avenue. *New York Academy of Medicine*, meets the first Wednesday of each month at the University.

Institution for the Blind, occupies a large and imposing Gothic edifice of granite, on Ninth avenue, in the north-west part of the city. Reached by the Ninth and Eighth avenue stages. Visitors received on Tuesdays, from 1 to 5 P. M. The Institution has about one hundred pupils.

Deaf and Dumb Asylum, Fiftieth street and Fourth avenue, via Harlem railroad. Visitors admitted from 1 to 4, P. M. The large and commodious building of this Institution accommodates about two hundred and fifty pupils.

The Bloomingdale Asylum for the Insane and the *New York Orphan Asylum*, are in the upper part of the Island, 7 miles from the City Hall, on the line of one of the pleasantest drives about New York.

Blackwell's Island. The City Penitentiary, the Lunatic Asylum, the Alms House, Hospital and Work House, on Blackwell's Island, in the East River, are worth the especial attention of the stranger. This Island, as also Ward's and Randall's Islands, may be reached by steamboat, from foot of Grand street, East River, at 12, M., daily; or by the Harlem stages to Sixty-first street.— Stages leave 25 Chatham street every fifteen minutes.

Ward's and Randall's Islands, near by, are occupied by the public charitable Institutions. The elegant and massive structures which cover this famous group of islands make a striking feature in the landscape, as we sail up the East River to the suburban villages on Long Island, or en route for Long Island Sound.

Churches. New York has nearly 300 churches, many of which are very costly and imposing edifices.—Among those most worthy the notice of the stranger, are *Trinity church* (Episcopal), in the lower part of Broadway; *St. Paul's* (Episcopal), not far off, in Broadway; *St. John's* (Episcopal), in St. John's Park; *St. Thomas'* (Episcopal), Broadway and Houston street; *Grace* church (Episcopal), Broadway and Tenth street; *Church of the Puritans*, Union sq.; *St. Paul's* (Methodist), Fourth avenue; *South Dutch*, Fifth avenue and Twenty-first street; *Dutch Reformed* (Dr. Hutton), Washington Square; *St. Mark's* (Episcopal, Dr. Anthon), Stuyvesant street; *St. George's* (Dr. Tyng, Episcopal), East Sixteenth street, Stuyvesant Square; *First Baptist*, corner of Broome and Elizabeth streets; *Amity Street*, Dr. Williams (Baptist), 31 Amity; *Madison Av.* (Baptist), Dr. Hague; *St. Patrick's Cathedral* (R. C.), corner of Prince and Mott streets; *Dutch Reformed*, Lafayette Place; *Dr. Potts'*, Presbyterian, in University Place, corner Tenth street; *Church of the Divine Unity* (Universalist), Dr. Chapin, 548 Broadway; *Church of the Messiah*, Unitarian, Dr. Osgood, 728 Broadway; *Church of All Souls* (Unitarian), Dr. Bellows, Fourth avenue, corner of Twentieth street; *Church of the Holy Redeemer*, German Catholic, Third street, near Avenue A; *First Presbyterian Church*, Dr. Phillips, Fifth avenue, between Eleventh and Twelfth streets; *Church of the Holy Communion*, Episcopal, Dr. Muhlenburg, Sixth avenue and Twentieth street; *Fifth Avenue Presbyterian* church, Dr. Alexander, corner of Nineteenth street; *French Church*, Protestant Episcopal, Dr. Verren, corner of Church and Franklin sreets; *Trinity Chapel*, Episcopal, 26th street, near Broadway; *Church of the Annunciation*, Episcopal, Dr. Seabury, Fourteenth street, between Sixth and Seventh avenues; *Church of the Ascension*, Episcopal, Dr. Bedell, Fifth avenue, corner of Tenth street; *St. Bartholomew's*, Episcopal, corner of Lafayette Place and Great Jones street; *Shaarai Tephita* (Gates of Prayer), Hebrew, 112 Wooster street; *Bnai Jeshurun* (Sons of Jeshurun), Hebrew, Greene street, near Houston.

THEATRES AND PLACES OF AMUSEMENT.

The Academy of Music, or Italian Opera House, is at the corner of Fourteenth street and Irving Place. Seats for 4,600 persons. This is one of the most sumptuous edifices of the kind in the world.

Niblo's Garden and Saloon, rear of Metropolitan Hotel, Broadway.

Winter Garden, one of the leading theatres, is in the rear of the Lafarge House, 641 Broadway.

Wallack's Theatre, another of the most popular establishments, is at 485 Broadway.

Laura Keene's Theatre is in Broadway, between Bleecker and Houston sts.

Palace Garden, a summer resort—14th st. between Fifth and Sixth avs.

Bowery Theatre, Bowery.

Barnum's Museum, corner of Broadway and Ann street.

MISCELLANEOUS OBJECTS AND PLACES WORTH SEEING IN THE CITY AND ON THE ISLAND OF NEW YORK.

Take either of the city railways (if time waits for employment), and explore, at will, the picturesque beauties of the upper part of the Island of New York—wander through the walks and woods of the Central Park—and yet miles beyond to the Harlem River and the villa-covered acclivities of Mount Washington, on the Hudson River side.

The Croton Aqueduct. *Receiving Reservoir* on York Hill, about five miles from the city; the *Distributing Reservoir* on Murray Hill, Fifth avenue, adjoining the Crystal Palace, and reached by the same routes and by the Fifth avenue stages.

The Croton Aqueduct, the greatest public work of the city, brings abundant supplies of water from the Croton Lake, 40 miles distant. The original cost of this magnificent labor was over thirteen millions of dollars.

High Bridge is a noble work constructed for the passage of the Croton Aqueduct over the Harlem River, from Westchester County to the Island of

New York. The Harlem railway to Harlem, six miles, and thence a short distance by omnibus.

The Arsenal is on Fifth avenue, between Sixty-second and Sixty-fifth streets, within the great area of the new Central Park. Harlem cars, or Sixty-first street stages.

The magnificent newspaper offices of the city, of the New York Herald, Tribune, Times and others, and the extensive book establishments of the Harpers and the Appletons, are places of exceeding interest.

Many of the private residences of the city, particularly the palatial abodes on the *Fifth avenue* and its vicinity, should be seen, if one would get any fair idea of the architectural beauty and splendor of the metropolis.

Harlem, a part of the city, at the north end of the Island, is upon the Harlem River. Cars from City Hall Harlem R. R. depots, seven miles.

Bloomingdale and **Manhattanville** are at the north end of the Island of New York.

THE ENVIRONS OF NEW YORK.

Places of interest in the vicinage of the City.

Hoboken and **Weehawken,** charming rural resorts—in summer-time—across the Hudson River, on the New Jersey shore. Here are delightful walks, for miles, along the margin of the river, on high ground overlooking the Bay and city, and all the country round—in shady woods and upon verdant lawns, and among wild forest glens. Ferry, every few minutes (fare three cents), from Barclay, Canal, Christopher and West Nineteenth streets.

Astoria, a suburban village on Long Island, six miles up the East River, near the famous whirlpool of Hell Gate, a place of beautiful villas. Steamboat, foot of Fulton street, East River, or by stage every hour, from 23 Chatham street, to foot of Eighty-sixth street—cross by Hell Gate Ferry.

Staten Island. *New Brighton, Port Richmond, and Sailors' Snug Harbor.* Ferry every hour and a half, from 8½,

A. M. to 6½, P. M., from foot of Whitehall street. To Quarantine, Stapleton and Vanderbilt's Landing, ferry every hour, foot of Whitehall street. Nothing can be more enjoyable than a sail down the Bay to any of the villages and landings of Staten Island; and nothing more agreeable than the sight of its many suburban villas, or of the superb views over land and sea which its high grounds command. Brighton is a particularly beautiful little village, with good hotels and boarding-houses. Near it, is the Sailors' Snug Harbor, an ancient foundation for dilapidated mariners. Two miles east of Brighton is the Marine hospital and the village of Tompkinsville, and its 3,000 people. The voyage to Staten Island occupies about half an hour.

Fort Hamilton, 8 miles down the Bay, commands, in connection with Forts Lafayette and Tompkins opposite, the passage seaward of the Narrows. A summer residence and resort for sea bathing. Via boats to Coney Island.

Coney Island, belonging to the town of Gravesend, is five miles long, and one broad, and is situated about 12 miles from New York. It has a fine *beach* fronting the ocean, and is much frequented, but not by ladies. On the north side of the island is an hotel. Steamboats ply regularly between the city and Coney Island during the summer season. *Fare,* 12½ cents, each way.

Rockaway Beach, a celebrated and fashionable watering-place, on the Atlantic sea-coast, is in a south-east direction from New York. The *Marine Pavilion,* a splendid establishment, erected in 1834, upon the beach, a short distance from the ocean, is furnished in a style befitting its object as a place of resort for gay and fashionable company. There is another hotel here which is well kept; also several private boarding-houses, where the visitor, seeking pleasure or health, may enjoy the invigorating ocean breeze, with less parade and at a more reasonable cost than at the hotels. The best route to Rockaway is by the Long Island Railroad to Jamaica, twelve miles, 25 cents; thence

Long Branch, New Jersey.

by stage eight miles, over an excellent road, to the beach, 50 cents.

Long Branch, situated on the eastern coast of New Jersey, 32 miles from the city of New York, is a popular place of resort for those fond of seashore recreation, and where a pure and invigorating atmosphere is always to be found.

The *Ocean House*, a hotel of the first order, a short distance north from Long Branch, is a place where, during the oppressive heat of summer, a greater degree of real comfort can be enjoyed than, perhaps, at the more fashionable watering-places.

There is admirable sport in this vicinity for the angler. The Shrewsbury river on the one side, and the ocean on the other, swarm with all the delicate varieties of fish with which our markets abound.

Shrewsbury, Red Bank, and **Tinton Falls**, in the vicinity of the above, are also places of great resort.

Flushing, on Long Island, 10 miles from the metropolis, is upon an arm of the Sound called Flushing Bay. The Linnæan Botanic Garden is here. Boat at Fulton street.

Flatbush, about five miles from Brooklyn, **Flatlands, Gravesend,** ten miles, are small but handsome places. Shores of the latter place abound with clams, oysters, and fowl, and are much resorted to.

Jamaica, another suburban town on Long Island, is 12 miles distant by the Long Island Railroad.

Greenwood Cemetery is in the south part of Brooklyn, at Gowanus, about three miles from the New York and Brooklyn ferries. One of the numerous railways which so thickly and so conveniently link all parts of Brooklyn, extends to the Cemetery ; the cars leaving the Fulton Ferry every five minutes, and speedily transporting the traveller for the small fare of five cents. On Sunday, only the owners of lots are admitted within the grounds.

This Cemetery was incorporated in 1838, and contains 242 acres of ground, about one half of which is covered with wood of a natural growth. It originally contained 172 acres, but recently 70 more have been added by purchase, and brought within the enclosure. Free entrance is allowed to persons on foot during week-days, but on the Sabbath none but the proprietors of lots and their families, and persons with them, are admitted; others than proprietors can obtain a permit for carriages on week-days. These grounds have a varied surface of hills, valleys, and plains. The elevations afford extensive views; that from Ocean Hill, near the western line, presents a wide range of the ocean, with a portion of Long Island. Battle Hill, in the northwest, commands an extensive view of the cities of Brooklyn and New York, the Hudson River, the noble bay, and of New Jersey and Staten Island. From the other elevated grounds in the Cemetery there are fine prospects. Greenwood is traversed by winding avenues and paths, which afford visitors an opportunity of seeing this extensive Cemetery, if sufficient time is taken for the purpose. Several of the monuments, original in their design, are very beautiful, and cannot fail to attract the notice of strangers. Those to the memory of Miss Canda, of the Indian Princess, Dohumme, and the "mad poet," Mc Donald Clark, near the Sylvan Water, are admirable; so also are the memorials to the Pilots and to the Firemen.

Visitors, by keeping the main avenue, called *The Tour*, as indicated by *guide-boards*, will obtain the best general view of the Cemetery, and will be able again to reach the entrance without difficulty. Unless this caution be observed, they may find themselves at a loss to discover their way out. By paying a little attention, however, to the grounds and guide-boards, they will soon be able to take other avenues, many of which pass through grounds of peculiar interest and beauty.

The New York Bay Cemetery is reached in a pleasant sail down the harbor. It is one of the most beautiful rural spots in all the beautiful vicinage of New York.

The U. S. Navy Yard is across the East River at Brooklyn. The *United States Naval Lyceum*, in the Navy Yard, is a literary institution, formed in 1833 by officers of the navy connected with the port. It contains a splendid collection of curiosities, and mineralogical and geological cabinets, with numerous other valuable and curious things worthy the inspection of the visitor. A *Dry Dock* has been constructed here at a cost of about $1,000,000. On the opposite side of the Wallabout, half a mile east of the Navy Yard, is the *Marine Hospital*, a fine building, erected on a commanding situation, and surrounded by upwards of 30 acres of well-cultivated ground.

At the Wallabout were stationed the Jersey and other prison-ships of the English during the Revolutionary war, in which it is said 11,500 American prisoners perished, from bad air, close confinement, and ill-treatment. In 1808, the bones of the sufferers, which had been washed out from the bank where they had been slightly buried, were collected, and deposited in 13 coffins, inscribed with the names of the 13 original States, and placed in a vault beneath a wooden building erected for the purpose, in Hudson avenue, opposite to Front street, near the Navy Yard.

The Atlantic Dock, about a mile below the South Ferry, Brooklyn, is a very extensive work, and worthy the attention of strangers. The Hamilton Avenue Ferry, near the Battery, lands its passengers close by. The company was incorporated in May, 1840, with a capital of $1,000,000. The basin within the piers contains 42½ acres, with sufficient depth of water for the largest ships. The piers are furnished with many spacious stone warehouses.

Jersey City, N. J., is on the Hudson, opposite the city of New York, with which it is connected by continual ferry, from foot of Cortlandt street (fare, four cents). In 1854, the population was about 21,000. Jersey City is the New York terminus of the Phila-

Brooklyn, N. Y.

delphia and New York and Erie Railroad routes, and of the Morris Canal. It is also the berth of the Cunard line of Atlantic steamers.

For *Newark* and other places near New York, upon the Philadelphia routes, see index. For suburban villages on the Hudson, see route from New York to Albany.

CITY OF BROOKLYN.

We have already spoken of many of the objects of interest in Brooklyn, in the preceding article upon New York; as the Navy Yard, Greenwood Cemetery, the Atlantic Dock, the neighboring Long Island villages of Astoria, Jamaica, Flushing, Rockaway Beach, etc. Besides these points, there is much else of interest across the river— many fine churches, and other public buildings.

Brooklyn possesses more than eighty church edifices of various denominations. Among the most costly and imposing are—

The Church of the Holy Trinity, Clinton street, Epis.

The Church of the Pilgrims, Congregational (Rev. R. S. Storrs).

Church of the Savior, Pierrepont st.,

cor. of Monroe, First Unitarian Congregational (Rev. F. A. Farley).

Grace Church, Brooklyn Heights, Epis. (Rev. Jared B. Flagg).

Christ Church, Clinton st (Dr. Canfield.

Plymouth Church, Orange street (Rev. Henry Ward Beecher).

First Reformed Dutch Church, Joralemon street (Dr. Dwight).

First Presbyterian Church, Henry, near Clark street.

Dutch Reformed Church, Pierrepont st. (Rev. Mr. Eells).

St. Ann's Church, Washington, near Sands street (Dr. Cutler).

Second Presbyterian Church, Fulton street, cor. of Clinton.

Hotels.—The *Pierrepont House* is an elegant establishment on the Heights ; the *Mansion House* is in the same eligible part of the city ; the *Globe Hotel* is at 244 Fulton street.

The *City Hall* (Court and Fulton streets), is one mile distant from the ferry. It is a handsome building of white marble, from the Westchester quarries. Its length is 162 feet, and its height to the top of the cupola is 153 feet. Cost, $2,000,000.

The *Post Office* is opposite the City Hall.

MAP OF THE

HUDSON RIVER,

AND

VICINITY

Showing the

RAILROADS.

0   5   10   15   20 Miles

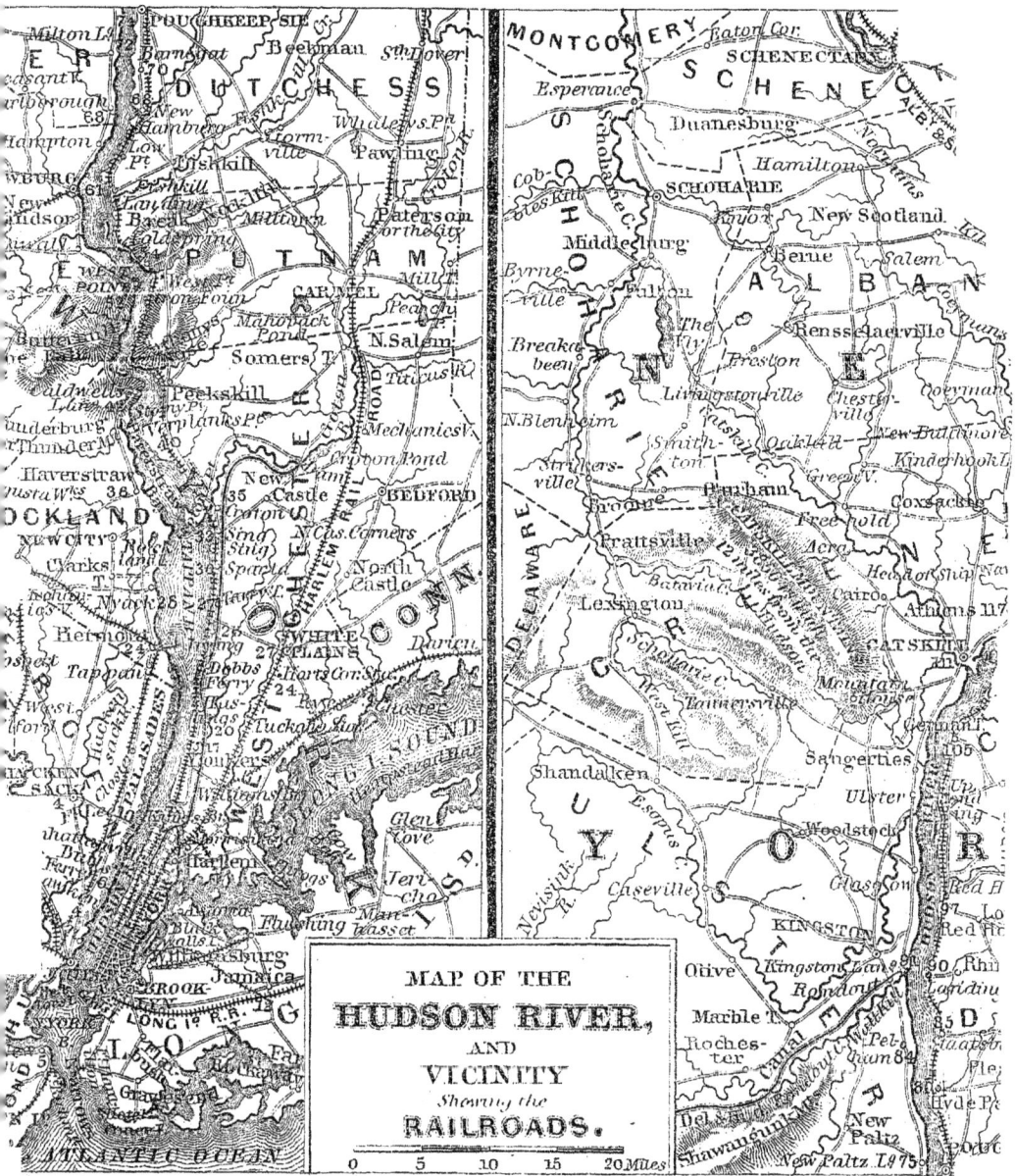

1847, by W. Williams, in the Clerk's Office, of the District Court, of the Sou.

DUTCHESS · PUTNAM · ROCKLAND · WESTCHESTER · CONN.
MONTGOMERY · SCHENECTADY · SCHOHARIE · ALBANY · GREENE · ULSTER

POUGHKEEPSIE · Barnegat · Beekman · S. Dover · Esperance · Duanesburg · Hamilton · Schenectady · Albany
Milton La. · Pleasant V. · Marlborough · Hampton · New Hamburg · Fishkill · Whaleys Pt. · Pawling · New Scotland · Salem
Newburg · New Hudson · West Point · CARMEL · Paterson or the City · Middleburg · Berne · Rensselaerville
Butter Hill · Mahopack Pond · N. Salem · Somers T. · Fulton · Preston · Chester ville · Coeyman
Haverstraw · Peekskill · Verplanks Pt. · Mechanicsv. · Breakabeen · Livingstonville · New Baltimore · Kinderhook
Caldwells · Sing Sing · Sparta · Croton · BEDFORD · N. Blenheim · Smithton · Oakland · Green V. · Coxsackie
ROCKLAND · New City · Clarks · Nyack · North Castle · Strykersville · Broome · Durham · Freehold · Acra · Head of Ship
Piermont · Tappan · Dobbs Ferry · WHITE PLAINS · Darien · Prattsville · Batavia · Cairo · Athens
WESTCHESTER · Chester · Lexington · Tannersville · CATSKILL · Mountain House · Germant.
Harlem · Jericho · Glen Cove · Manhasset · Shandalken · Woodstock · Sangerties · Ulster
Williamsburg · Jamaica · BROOKLYN · LONG IS. R.R. · Caseville · Glasgow · Red H.
NEW YORK · Flushing · Rockaway · Olive · KINGSTON · Kingston Land. · Rondout · Rhin.
ATLANTIC OCEAN · Marble T. · Rochester · New Paltz · Shawangunk · Hyde P.

LONG ISLAND SOUND · HUDSON · PALISADES · DELAWARE CO. · GREENE CO. · ULSTER CO.

The *Brooklyn Athenæum*, corner of Atlantic and Clinton streets, in South Brooklyn, is a fine edifice of brick, with brown stone facings. It has an admirable library, reading-rooms, and a spacious lecture or concert hall.

The *Lyceum*, containing the city Library and a good lecture-room, is at the corner of Washington and Concord sts.

The new *Academy of Music*—a superb structure, is in Montague street.

Brooklyn, which now comprehends also the city of Williamsburg, is in population (which is no less than 200,000) the second city in the State of New York, though its close vicinage to the metropolis absorbs it, and destroys its distinctive importance. A great portion of its residents do business in New York, and live in Brooklyn only for the convenience and comfort of purer air, more quiet, and less cost.

The city is in many parts elegantly built, and the bold position on the Heights, directly looking down upon the river and the bay, is a charming site for a summer abode. Some of the avenues of Brooklyn are wide, and delightfully lined with cottage residences.

The numerous ferries across the East river afford pleasant and perpetual access to Brooklyn.

Fulton Ferry—From Fulton st., N.Y., to Fulton st., Brooklyn, every five minutes in the day time, at a fare of *two cents.*

South Ferry—From Whitehall street, N. Y., to Atlantic st., Brooklyn.

Hamilton Ferry—Whitehall st., N.Y., to Hamilton avenue and Atlantic Docks, Brooklyn.

Catharine Ferry—Catharine st., N.Y., to Main st., Brooklyn.

Jackson Ferry—From Gouverneur st., N. Y., to Jackson st., Brooklyn.

Wall Street Ferry—Wall st., N. Y., to Montague st., Brooklyn.

Roosevelt Ferry—Roosevelt st., N.Y., to Bridge st., Brooklyn.

To East Brooklyn, or Williamsburg.

Peck Slip, Grand street, and Houston street, N. Y., every ten minutes.

NEW YORK TO ALBANY AND TROY.

It is fortunate for the gratification and the cultivation of the public taste for the sublime and beautiful in natural scenery, when our great highways of travel chance to lead through such wondrous landscape, as does our present journey up the Hudson River, from New York to Albany. Even to the wearied or the hurried traveller this voyage is ever one of pleasure, in its unique and constantly varying attractions, its thousand associations, legendary, historical, poetical, and social.

The Hudson received its name in honor of Hendrick Hudson, a Dutch navigator, who discovered it and ascended its waters for the first time, in 1607, in his good barque, the Half-Moon. It is also known as the North River, which name was given to it by the original Dutch colonists, to distinguish it from the South (Zuyd), as they called the neighboring floods of the Delaware. Its source is in the mountain region of the Adirondack, in the upper portions of New York, whence it flows in two small streams, the one from Hamilton, and the other from Essex County. These waters, after a journey of 40 miles, unite in Warren County. The course of the Hudson varies from south by east to east for some distance, but at length drops into a straight line, and continues thus, nearly southward, until it falls into the Bay of New York. Its entire extent is about 300 miles; its navigable length, from the sea to Albany, is half that distance. Its breadth, near the head of steamboat navigation, varies from 300 to 900 yards; and, at the Tappan Bay, 20 miles above the city of New York, it widens to the extent of from four to five miles. Ships of the first class may ascend to Hudson, a distance of 117 miles, and small sailing craft may reach the head of tide water (166 miles), at Troy. The number of steamboats and other vessels upon the river may be counted by thousands.

To the Hudson belongs the honor, not only of possessing the finest river steamers in the world, but of *having borne upon its waters the first steamboat which ever floated,* when Robert Fulton ascended the river in the Clermont, in 1807, exactly two centuries after the first voyage of Hendrick Hudson in the Half-Moon.

Every possible facility is now at command for the passage of the Hudson, either by steamer or by railway, morning, noon, and night. The commercial traveller, thinking more of his destination than of the pleasures by the way, will take the railroad route, while the pleasure-seeking tourist, in quest of the picturesque, and with time to enjoy it, will assuredly go by water:

RAILWAY ROUTE.

The journey by the Hudson River Railway, 144 miles, to Albany, is a poem in prose. The road lies on the eastern bank of the river, kissing its waters continually, and ever and anon crossing wide bays and the mouths of tributary streams. Incredible difficulties have been surmounted in its mountain, rock, and water passage, and all so successfully and so thoroughly, that it is one of the securest railway routes in the world. With all its immense business, its history is happily free from any considerable record of collision or accident whatever. This is owing as much to the vigilant management, and the admirable police, as to the substantial nature of the road. Flag-men are so stationed along the entire line, at intervals of a mile, and at curves and acclivities, as to secure unbroken signal communication from one end to the other.

Trains leave Chambers street and College Place almost hourly. Fare, usually, $3. Time, about five hours.

STATIONS.

(For description of places and scenes, see Steamboat Route following.)

Chambers street, New York; Thirty-first street, New York; Manhattan, 8 miles from New York; Yonkers, 17; Dobb's Ferry, 22 (Ferry to Piermont, Erie Railway); Tarry Town, 27; Sing Sing, 32; PEEKSKILL, 43; Garrison's, 51 (Steam Ferry to West Point and Cozens' Hotel); Cold Spring, 54; Fishkill, 60 (will be the junction of Providence, Hartford, and Fishkill Railroad, Steam Ferry to Newburgh, terminus of Newburgh Branch of Erie Railway); New Hamburgh, 66; POUGHKEEPSIE, 75 (Half-way and refreshment station); Hyde Park, 81; Staatsburg, 85; Rhinebeck, 91; Barrytown, Tivoli, 100; Germantown, 105; Oakhill, 110 (Ferry to Catskill village, route to Catskill Mountains); HUDSON, 116 (Railway route to Boston, via Hudson and Berkshire road); Stockport, 120; Coxsackie, 125; Stuyvesant, 126; Schodack, 133; Castleton, 136; East Albany, 144 (Ferry to Albany); Troy, 152 miles.

STEAMBOAT ROUTE.

If the traveller accompany us up the Hudson, he will take passage in one of the noble steamers (very fittingly called floating palaces), which leave New York every morning and night.

The size and beauty of the boats, and the conveniences, comfort, and luxury of all their appointments, will be matter for pleasant wonder and thought, even to those most accustomed to them, whenever a moment can be stolen from the endless attractions of the way.

We start as the morning sun is falling upon the thousand sail which fill the grand Bay of New York; but scarcely have our eyes taken in half the beauties of this superb panorama—the roofs, and spires, and domes of the great metrop-

* Not taking into the account the various previous approximations to this great result, as the unsuccessful attempt in England (1737) by Jonathan Hulls; that of Mr. Miller, of Dalswinton, in Dumfrieshire, 1776; another in 1786; the Venture, in 1794, of the Earl of Stanhope; Mr. Henry Bell's model, made in 1800, for Lord Viscount Melville, then at the head of the Board of Admiralty; and the somewhat more successful trial, in 1801, of Mr. Symington. This gentleman constructed a steamvessel on the Forth and Clyde Canal, which went, with ease, at the rate of four miles an hour; but the agitation produced by it was so great that it was feared the banks of the canal would be injured, and the idea was, therefore, abandoned. To Robert Fulton, however, belongs, without any controversy, the honor of having first applied steam navigation to any practical and useful purpose.

The Palisades on the Hudson.

olis on one side, Jersey City upon the opposite shore, the fortresses of Governor's Island, of Bedloe's, of Ellis's Islands, and of Fort Hamilton ; the shores of old Long Island, and the villa banks of Staten Island beyond, with the far-off perspective of the hill-bound "Narrows"—before we must turn our backs upon it all, to gaze upon the yet more charming scenes which are presented to us as our steamer turns its prow northward.

Along we sail, past the streets and wharves of the city, which seem interminable in succession, but our eyes fall upon the wooded shores at last, upon the elegant country villas peeping out from among the trees on the one hand, and the tall cliffs of the far-reaching Palisades on the other. The wilderness of brick and stone is behind us and forgotten, in the presence of green fields and rustling woods. Even the suburban charms of Hoboken, and the precipices of Weehawken, with its grave and memories of the unfortunate Hamilton, give place, in our esteem, to the more rural landscape upon which we now

enter. Let us peep as closely as our rapid flight may permit at each passing village, city, and scene. First come

The Palisades. These grand precipices, rising to the height of 500 feet, follow, in unbroken line, as far as that great bay of the river called the Tappan Sea, a distance of 20 miles. The rock is trap, columnar in formation, something after the fashion of the famous Giant's Causeway and of Fingal's Cave in Ireland. They lend great beauty to the picture as we start upon our journey, and to all the pictures of the river, into which they come.

Bull's Ferry, six miles from the city, now lies upon our left. It is a favorite summer resort and residence of the people of New York. In the hot months, the ferry boats, continually plying thither, at a fare of only 12½ cents, are ever well freighted with merry passengers.

Bloomingdale, a suburban village five miles from the City Hall is now upon our right. The Orphan Asylum here, with its emerald lawns, sloping down to the

6

quiet waters, is a pleasant picture for both eye and heart.

Fort Lee, ten miles up the river, and opposite 160th street, New York, now calls us back again to the western shore. It crowns the lofty brows of the Palisades. Some interesting memories of the days of the American Revolution are awakened here. The anxious thoughts of Washington and his generals turned to this point in that eventful period. A fortification here stood upon the heights, which was called Mount Constitution, and here it was attempted, by the express command of Congress, to obstruct the navigation of the river by every art, and at whatever expense, " as well to prevent the regress of the enemy's frigates lately gone up as to hinder them from receiving succors." A large force of Americans, in retreating from Fort Lee, were overpowered, and either slain or taken prisoners by a greatly superior body of Hessian troops.

Fort Washington, another spot of deep historical interest, lies nearly opposite to Fort Lee, and, like that locality, reminds us of the most trying hours of the trying times in American story. It fell into the hands of the enemy, November 16th, 1776, and the garrison of 3000 men became prisoners of war. Two days after, November 18th, Lord Cornwallis, with 6,000 men, crossed the river, at Dobb's Ferry, and attacked Fort Lee. The garrison there, then commanded by General Greene, made a hasty retreat to the encampment of the main army, under Washington, five miles back, at Hackensack. All the baggage and stores fell into the hands of the enemy. Had the English general followed up his successes at this period, with proper celerity and energy, he would most likely have effectually crippled the American army. Fort Washington is situated upon the highest part of Manhattan Island, between 181st and 186th streets, New York. It is between 10 and 11 miles from the City Hall. The fort was a strong earthwork of irregular form, covering several acres. Some 20 heavy cannons, besides smaller arms, bristled upon its walls, though its strength lay chiefly in its position. The very spot where the old fort once stood, as well as all the region round, is now covered by the peaceful and fragrant lawns and gardens of elegant villa residences. Just below the high ground once occupied by Fort Washington, and close by the river, is the promontory of Jeffrey's Hook: a redoubt was constructed here as a covering to the *chevaux-de-frise* in the channel. The banks of this work are still plainly to be seen. Above Fort Washington, and still upon the eastern side of the river, was Fort Tryon. This site now lies between 195th and 198th streets, New York. Not far beyond, is the northern boundary of Manhattan Island—the little waters, famous in history and story, as Spuyten Duyvel Creek (Spite the Devil). Hard by (217th street) was a redoubt of two guns called Cock Hill Fort; and upon Tetard's Hill, across the creek, was Fort Independence, a square redoubt with bastions. There was still another military work here, strengthened by the British in 1781, and named Fort Prince. The upper end of the island of New York, where we have lingered so long, is rich in scenes and memories of interest; and the beautiful landscape is yet embellished by abundant traces of all its ancient history.

Yonkers, 17 miles up the river, is an ancient settlement at the mouth of the Neperan, or Saw-Mill River. Since the opening of the railway, it has become a fashionable suburban town of New York, as the short distance thence permits pleasant, and speedy, and cheap transport by land or water.

Yonkers was the home of the once famous family of the Phillippses, of which was Mary Phillippse, the first-love of General Washington. The family exercised manorial rule in the neighborhood, and their ancient mansion is still to be seen.

East of the manor-house of the Phillippses, is Locust Hill, where the American troops were encamped, in 1781. Near the village is the spot where Colonel Gist was attacked (1778) by a combined force under Tarleton and others. In 1777, a naval action occurred in front of Yonkers, between the American gun-boats and the British frigates, Ross and Phœnix.

Mr. Frederic Cozzens, the writer, resides at Yonkers, and some pleasant reminiscences of his home may be found in his genial "Sparrowgrass" papers.

Fonthill—Academy of Mount St. Vincent. The "Castle" of Mr. Edwin Forrest, known as Fonthill, is just below Yonkers. It is now, together with a larger and more imposing edifice, owned and occupied by the R. C. School of Mt. St. Vincent.

Hastings, three miles north of Yonkers, is a thriving little village, and its fortunes are daily improving with the favors of the citizens of New York, who eagerly seek homes amidst its pleasant and healthful places. Some of the country seats in the neighborhood— and they are numerous—are very elegant and luxurious establishments.

Dobb's Ferry, two miles yet beyond, and still upon the eastern bank of the river, is an ancient settlement, with a new leaven of metropolitan life, like all the places within an hour or two's journey from New York. The village has a pleasant air, lying along the river slope, at the mouth of the Wysquaqua Creek. Its name is that of an old family which once possessed the region and established a ferry. We are led back again here, to the times of the Revolution, and especially to that dramatic episode—some of the scenes of which transpired here and hereabouts—the story of Arnold and André. Remains of military work still exist at Dobb's Ferry.

Irvington and "Sunnyside." Irvington, to which we now come, still on the right or eastern bank, was once called Dearman, and it was expected to grow into a large town, as an outlet of the Great Erie Railway, which touches the river opposite, at Piermont; but the Erie travel was afterwards led to the metropolis through another terminus at Jersey City, and so

Irvington is little more than a railway station to this day.

Dearman was rechristened Irvington in honor of the late beloved author, Washington Irving, whose unique little cottage of SUNNYSIDE is close by upon the margin of the river, hidden from the eye of the traveller only by the dense growth of the surrounding trees and shrubbery.

Piermont is on the left or western bank of the widest part of the Hudson, called the Tappan Bay or Sea, in the heart of which we are now sailing. It was born of, and has grown up from, the business of the Erie Railway, of which it is a terminus, and was once the only eastern terminus, the route of the road having originally been entirely continued, as it is now in part, thence down the Hudson to New York. The river here is three miles in width, and the shores, particularly upon the west, are so varied and bold, as to present most striking and attractive pictures. Piermont, rising from the water's edge to the villa-crowned summits of lofty hills, and with its grand railway pier reaching out a mile or more into the river, is not one of the least pleasing features of the scenery of the Tappan Bay.

Mr. Lewis Gaylord Clark, editor of the "Knickerbocker Magazine," lives upon the eminence here, in a little house, which he calls "CEDAR HILL COTTAGE."

Two or three miles back of Piermont is the old town of Tappan, interesting as having been one of the chief of Washington's head-quarters during the Revolution ; and as the spot also where Major André was imprisoned and executed. The home of the commander-in-chief and the jail of the ill-fated officer are still in good preservation, though the latter house has been somewhat modified in its interior arrangements of late years, to suit its present occupancy as a tavern, under the style and title of the "'76 Stone House." The old Dutch church, in which André was tried, stood near by, but it was torn down in 1836, and a new structure reared upon its site. The spot where the execution took place (October 2d,

1780) is within a little walk of the old Stone House, in which the prisoner was confined.

Nyack is the next village above Piermont, and upon the same side of the river, while Tarrytown lies directly opposite it, and is connected therewith, and with New York, by a steam ferry. Beds of red sandstone were once industriously quarried at Nyack.

Tarrytown is a very active, prosperous little place on the eastern bank of the Hudson. It has many attractions, historical, pictorial, and social. Elegant villas, chiefly occupied by New York gentlemen, have gathered thickly around it, as about all this part of the river's marge, within the past few years ; among them is Mr. Irving's homestead of Sunnyside, at Irvington, two miles below, and a mile or more distant, in the opposite direction, is the quiet little valley of Sleepy Hollow, which he has wreathed with such a garland of poetic remembrances and fancy, through his charming legends and tales. The visitor at Tarrytown will neglect many things before he denies himself the pleasure of a stroll to Sleepy Hollow, where Diedrich Knickerbocker roamed and meditated in days gone by ; and of a walk by the Pocantico, and across the bridge, over which Ichabod Crane was pursued by the Headless Horseman. The scenes are all there still ; and so the old Dutch church, to which the luckless pedagogue fled for sanctuary.

During the Revolution, Tarrytown witnessed many stormy fights between those lawless marauding bands of both British and Americans, known as Skinners and Cow-boys. The ground suited their wants, as it lay between the encampments of the two armies, and was in possession of neither. It was upon a spot, now in the heart of Tarrytown, that Major André was arrested, while returning to the British lines, after a visit to General Arnold. A simple monument—an obelisk of granite—now occupies the ground.

Sing Sing, on the right as we ascend, is 33 miles from the city. In its acclivitous topography, upon a hill-slope of 200 feet, it makes a fine appearance from the water. The greatest breadth of the Hudson, nearly four miles, is at this point. Many fine country seats crown the heights of this pleasant village. It is distinguished for its educational establishments ; for its vicinage to the mouth of the Croton river, from whence the city of New York derives its abundant supply of water ; and for being the seat of the chief prison of the State.

The Croton enters the Hudson two miles above the village, where its artificial passage to the metropolis is begun. The great aqueduct at this point is especially interesting, being carried over the Sing Sing Kill by an arch of stone masonry 88 feet between the abutments, and 100 feet above the water. The State Prison, which no visitor will fail to see, is located on the bank of the Hudson, nearly three quarters of a mile south of the village. The buildings are large structures, erected by the convicts themselves, with material from the marble and limestone quarries which abound here, and which many of them are continually. employed in working. The prisons form three sides of a square. The main edifice is 484 feet long, 44 feet wide, and 5 stories high, with cells for 1,000 occupants, 869 of which were filled in 1852. The female prisoners are lodged in a fine edifice, some 30 or 40 rods east of the male departments. The prisons are guarded by sentinels, instead of being enclosed by walls. The whole area covered by the establishment is 130 acres of ground. The railway passes through and beneath the prisons, but from the river they are all seen to advantage. The convicts not employed in working the marble quarries are engaged in the pursuit of various mechanical arts and trades. Sing Sing is a bustling business town. Its population over 3,000. Though the river communication with New York is not so great since as before the building of the railroad, way-steamboats from the city yet touch here daily.

Verdritege's Hook, opposite Sing

Sing, is a commanding height, with such a deceptive appearance from the water above and below of a grand headland, that it has been christened Point-no-Point. Coming near it, its promontory look entirely disappears, and it proves to be only, as we once called it elsewhere, a topographical will-o'-the-wisp.

Upon this mountain summit there is a charming pellucid water called **Rock-land Lake.** It is about four miles in circumference, and forms the source of the Hackensack River. Though not more than a mile from the Hudson, it is yet 250 feet above it. It is from this crystal lake that New York gets its best supplies of ice, which is cut into large square blocks. These blocks are then slid down to the level of the river, and when the winter passes they are transported to the city. Every voyager will bestow a pleasant thought upon the Rockland Lake, as he passes, in gratitude for the cooling beverages it gives him in the hot summer months, be that beverage julep, cobbler, cocktail, or Croton.

Haverstraw is also on the west side of the river, 36 miles up. It is a pleasant and prosperous place, of much picturesque vicinage. Some charming brooks, upon which artists delight to study, come into the Hudson here. We touch now again upon sacred ground, as we reënter amidst the scenes of our Revolutionary history; for directly opposite is Verplanck's Point, and in the immediate vicinity is the famous battle-ground of Stony Point.

Verplanck's Point, on the east side of the Hudson, is the spot at which Hendrick Hudson's ship, the Half-Moon, first came to anchor, after leaving the mouth of the river. Great was the marvel and terror of the astonished natives at that extraordinary event. "Filled with wonder," says Lossing, "they came flocking to the ship in boats, but their curiosity ended in a tragedy. One of them, overcome by acquisitiveness, crawled up the rudder, entered the cabin window, and stole a pillow and a few articles of wearing apparel. The mate saw the thief pulling his bark for land, and shot at and killed him. The ship's boat was sent for the stolen articles, and when one of the natives, who had leaped into the water, caught hold of the side of the shallop, his hand was cut off by a sword, and he was drowned. This was the first blood shed by these voyagers. Intelligence of it spread over the country, and the Indians hated the white man ever after." The creek which winds through the marsh, south of Verplanck's Point, as, afterwards, the peninsula itself, was called Meahagh by the Indians. Stephen Van Cortlandt purchased it of them in 1683, and it passed from his possession into that of his son, whose only daughter and heiress married Philip Verplanck, from whom its present name. Topographically, Verplanck's Point may be described as a peninsula, gradually rising from a gentle surface, until it terminates in the river in a bold bluff of from 40 to 50 feet elevation. The railway recedes here from the river-shore, and takes a seemingly inland route across the neck of the peninsula. "Here," says Mr. Lossing, in his Field-Book, from which we have already quoted, "during the memorable season of land and town speculation, when the water-lot mania emulated that of the tulip and the South Sea games, a large village was mapped out, and one or two fine mansions were erected. The bubble burst, and many fertile acres there, where corn and potatoes once yielded a profit to the cultivator, are scarred and made barren by intersecting streets, not *de*-populated but *un*-populated, save by the beetle and the grasshopper."

The narrowness of the river between this bluff and the opposite promontory of Stony Point, makes it the lower gateway of the Hudson Highlands, and renders it easily defensible against any possible hostile force. A small fortification, called Fort Fayette, once existed at the western extremity of Verplanck's Point, many remains of which are yet distinctly visible. This fort, and that of Stony Point opposite, were taken by the English, under Sir Henry Clinton, June

1, 1779. The garrisons at the time consisted, respectively, of only 70 and 40 men. Sir Henry Clinton immediately proceeded to strengthen his new possessions, while Washington was meditating their recapture, as the passage which they controlled was important to the free communication between the ·northern and southern portions of his army. We must now look across to

Stony Point.—The old lighthouse here calls this scene loudly to the notice of all passers. The beacon is placed amidst the remains of the ancient fort, and exactly upon the former site of the magazine. As we have said, the fort here, together with that upon Verplanck's Point opposite, fell into the hands of the enemy on the 1st of June, 1779. Despite its natural defences, and the additional strength which the enemy industriously gave it, the Americans determined to regain their lost possession. General Wayne, who was to command the proposed assault, is reported to have said to Washington, with daring emphasis, apropos of the dangers before him in this perilous venture: " General, I'll storm hell, if *you* will only plan it !" He did storm Stony Point on the night of July 15th, 1779, and next day he wrote to the commander-in-chief that the fort and garrison were his! It was a gallant exploit, and we wish we had the opportunity to review the whole story; but there are many miles yet between us and Albany, and we must move on to

Peekskill.—We now enter Haverstraw Bay, the second of the great extensions of the Hudson, and the commencement of the magnificent scenery of the Highlands. On our left is the rugged front of the Dunderburg Mountain, at whose base the little hamlet and landing of Caldwell are nestled ; on the right, the village of Peekskill ascends from the shore to the lofty hill summit, and before us is the narrow passage of the river, around the point of the Dunderburg, the grand base of Anthony's Nose, and other mountain cliffs and precipices. Let us look a moment, before we pass on, at Peekskill and its memories. The village was named after John Peek, one of the early Dutch navigators, who mistook the creek which comes into the river just above for the continuation of the Hudson itself (not an unreasonable mistake, so uncertain seems its direction at this highland pass), and thus thinking himself at the end of his journey, ran his craft ashore, and commenced his settlement. The present village was first settled in 1764. Its position is ex-

A View on the Hudson River.

ceedingly picturesque. A romantic brook comes down a deep glen, in the centre of the town, as it descends from the elevated plateau to the river, disfigured not a little at this day by the houses and foundries near it.

Noble views may be found everywhere here, and in every direction of the river and the surrounding country. From Gallows Hill northward (so called in remembrance of the execution there of a spy in the days of the Revolution), a grand panorama is exposed. Here, to the west, overlooking the village, the river, and its mountain shores ; there, southward, hill and valley, as far as the high grounds of Tarrytown below; and above, the Canopus valley, in the shadow of the Highland precipices. The division of the American army under Putnam, in 1777, was encamped upon Gallows Hill. Beneath this lofty ground, and upon the banks of Canopus Creek, is Continental Village, which was destroyed by General Tryon (Oct. 9, 1777), together with the barracks, capable of accommodating 2,000 men, and also much public store and many cattle.

The Van Cortlandt House, in this vicinage, is an object of interest, as the ancient seat of an ancient family, and as the temporary residence of Washington. Near by is a venerable church, erected in 1767, within whose graveyard there is a monument to the memory of John Paulding, one of the captors of Major André.

At the date of the first edition of this work (1857) there stood, in the streets of Peekskill, another of those venerable roofs, sacred as having at one time sheltered the head of the American chieftain. But, alas! it has since passed away, as are fast passing all its fellow-shrines. Thus fewer and fewer is the number of such spots to which we shall be able to direct the traveller as the years speed on.

A pleasant ride from Peekskill is to Lake Mahopac, a fashionable summer resort for the pleasure-seekers of New York. See Index.

The population of Peekskill in 1854 was 2,500. It is 43 miles from the city of New York, by rail.

Caldwell's Landing, at the foot of Dunderburg Mountain, was a calling-place for the river steamers, when the chief travel was by water, instead of by rail, as at the present day. The passengers for Peekskill, opposite, were then always landed at Caldwell. This spot is memorable for the search so seriously and actively made for the pirate treasure which the famous Captain Kidd was supposed to have secreted at the bottom of the river here. Remains of the apparatus used for this purpose are still seen, in bold, black relief, at the Dunderburg point, as the boat rounds it, towards the Horse Race. This Quixotic exploration has at least proved to a certainty that much valuable treasure *now* lies buried here, however uncertain the matter was before!

The Highlands. This grand mountain group, through which the Hudson now makes its way, extends from northeast to south-west, over an area of about 16 by 25 miles. The landscape which these noble heights and their picturesque and changeful forms present, is of unrivalled magnificence and beauty, whether seen from their rugged summits, or from the river gorges.

Thus says Theodore Fay of these scenes —

" By wooded bluff we steal, by leaning lawn,
  By palace, village, cot, a sweet surprise
  At every turn the vision breaks upon,
  Till to our wondering and uplifted eyes
  The Highland rocks and hills in solemn grandeur rise."

" Nor clouds in heaven, nor billows in the deep
  More graceful shapes did ever heave or roll;
  Nor came such pictures to a painter's sleep,
  Nor beamed such visions on a poet's soul!
  The pent-up flood, impatient of control,
  In ages past here broke its granite bound,
  Then to the sea in broad meanders stole,
  While ponderous ruin strew'd the broken ground,
  And these gigantic hills for ever closed around."

This powerful river, says another writer, writhes through the Highlands in abrupt curves, reminding one when the tide runs strongly down, of Laocoon in the enlacing folds of the serpent. The different spurs of mountain ranges which meet here abut upon the river in bold precipices, from five to fifteen hundred feet from the water's edge; the foliage hangs to them from base to summit, with the tenacity and bright verdure of moss; and the stream below, deprived of the slant lights which brighten its depths elsewhere, flows on with a sombre and dark green shadow in its bosom, as if frowning at the narrow gorge into which its broad-breasted waters are driven.

Passing round the point of Dunderburg (or Donderbarrack, the *Thunder Chamber*) we enter the swift channel called the Horse Race. On our right, in this wild and narrow gorge of the giant hills, are the rugged flanks of **Anthony's Nose,** bold rocky acclivities which rise to the height of 1,128 feet above the water. Two miles above is the **Sugar Loaf** mountain, with an elevation of 806 feet. Near by, and reaching far out into the river, is a sandy bluff, on which Fort Independence once stood. Further on is Beveridge Island, and in the extreme distance, Bear Mountain. Forts Clinton and Montgomery, taken by the British troops, after traversing the Dunderburg mountain, are in this vicinity; and so, too, a little lake called Skinnipink, or Bloody Pond, where a disastrous skirmish occurred on the eve of the capture of the forts and the consequent opening to the enemy of the passage of the Highlands. On this (the west) side of the river the **Buttermilk Falls** are seen descending over inclined ledges, a distance of 100 feet. They form a pleasant passage in the river landscape, though in themselves they are not especially picturesque.

In the heart of the Highland Pass and just below West Point on the west bank is **Cozzens'**—a spacious and elegant summer hotel, which comes most charmingly into the pictures of the vicinage. It is accessible, as is West Point at the same time, from the railway on the opposite side of the river, by a steam ferry from **Garrison's Station,** between Peekskill below and Cold Spring above. The concourse of

sail sometimes wind-locked in the angles of this mountain pass, is a wonderful sight. "This channel," says Mr. Willis, "is narrow and serpentine, the breeze baffling, and small room to beat; but the little craft will work merrily and well; and dodging about, as if to escape some invisible imp in the air, they gain point after point, till at last they get the Donderbarrack behind them, and fall once more into the regular current of the wind."

Constitution Island, with the rocky plateau of West Point, now bars our view of the upper portion of the Highland passage. Rounding it, we come into that wonderful reach of the river, flanked on the west by the royal cliffs of Cronest and Butter Hill, or Storm King, and upon the east by the jagged acclivities of Breakneck and Bull Hill, with the pretty village of Cold Spring beneath. From the heights of West Point delicious views of this new chapter of the river beauties may be obtained. Constitution Island was called, prior to the Revolution, Martelaer's Rock. It was fortified together with West Point, hard by, in 1775–6, when **Fort Constitution** was built, the remains of which still exist. Those of the magazine especially are well preserved, on the highest point, near the western extremity of the bluff. From this Island to West Point a chain was thrown across the river as an obstruction to the enemy's ships. Some links of this defence are to be seen in the neighborhood.

West Point, both from the unrivalled charms of its scenery and from its position as the seat of our most famous military school, is one of the most attractive spots upon the Hudson. It is replete with interest, too, as the centre of the important interests and incidents which in the days of the Revolution wove such a web of story and romance about all this portion of the beautiful river.

The edifices of the United States Military Academy, in full view as we approach, occupy a noble plateau, about a mile in circuit, and 188 feet above

The Highlands—West Point—The Beverly House—Cronest.

the water; and grand hills, which were fortified in the war time, leaving at this day romantic ruins to embellish the landscape, rise hundreds of feet yet above. It was the same bold and varied physical aspect of this spot which now delights the lover of nature's wonders, that in other days gave it its grand value, and its memorable fame as a site of military operations and achievements. The visitor will delight his eye at all points, whether he gaze upon the superb panorama of the river as he sits upon the piazza of the hotel upon the plateau, or whether he looks upon the scene from the yet loftier eminence above, crowned by the ruins of ancient fortresses; or whether he stroll amidst the interlacing walks, with new vistas of beauty and fresh memories of a gallant gone-by at every turn and step. When the remains of the old forts Putnam, Clinton, Webb and Wylly's have been seen, together with the little glen below the Parade Ground, called " Kosciusko's Garden," and embellished with an obelisk erected to the honor of the gallant Pole, the visitor will be ready to explore the edifices of the Academy establishment, and the many objects of interest which they contain; among them, Revolutionary relics and cannon captured in the war with Mexico. If his visit be in the months of July or August, the pleasures of the place will be agreeably increased by the picturesque scene of the annual encampment, on the broad terrace, of the Cadets, and by the daily practice of the military band. If he can gain the entrée of the studio of the distinguished painter Weir, who resides here, as Professor of Drawing in the Academy, he will be fortunate.

The United States Military Academy at West Point was established by Congress in 1802, and it is entirely controlled and supported by the Government.— The education of the Cadets is gratuitous, but each one is required to spend eight years in the public service, unless he be sooner excused. The course of study lasts five years, and embraces every theme required for a thorough

mastery of the military art. The graduates number more than 3,000.

West Point, in the Revolution, was the great key of the river, which Arnold, then in command of the post, would have betrayed into the possession of the enemy, but for the providential arrest of his co-plotter, André, at Tarrytown below.

The Robinson or Beverly House, occupied by Arnold at the time of his meditated treason, at which he received intelligence of the arrest of Major André, and from whence he made his escape to the British vessel, the Vulture, lying near by in the river, is on the opposite (east) bank, a pleasant drive of four or five miles south from Cold Spring. It is situated upon a fertile meadow at the foot of the Sugar Loaf Mountain—the lofty elevation on the east, which proves so Protean in form—now a bold cone, and now a ridgy line, as seen from below or from above. This homestead is now occupied by Lieutenant Thomas Arden, and is called " Ardenia." It has been kept in thorough repair, and its old aspect has been always religiously preserved.

Cronest casts its broad evening shadow upon us as we continue our voyage up from West Point. This is one of the grandest mountains found in the Highland group. Its height is 1,428 feet. From the summit, which may be readily reached, wonderful pictures of far and near are exposed to view.

The poet, George P. Morris, has happily sung the beauties of these bold cliffs—

" Where Hudson's waves o'er silvery sands
  Wind through the hills afar,
  And Cronest, like a monarch stands,
  Crown'd with a single star! "

The tourist, as he passes this romantic ground, will not fail to recall the scenes and incidents of Drake's charming story of the Culprit Fay, with its classic whispers of the dainty Fairy doings here.

Butter Hill, or Storm King, as Mr. N. P. Willis has re-named it, is the next mountain crest, and the last of the Highland range upon the west. The

6*

130   NEW YORK.

The Hudson—Cold Spring and " Undercliff"—Cornwall and " Idlewild "—New Windsor.

jealous people on the opposite shore, say that Butter Hill is only a corruption of But-a-hill! It would, though, be irreverent to believe in this derivation, for the Storm King, with its 1,500 and more feet of bold cliff and crag, is not an object to be spoken, or thought, lightly of.

Between Cronest and Storm King, (if we may adopt Mr. Willis's nomenclature) and in the laps of both, is a lovely valley, replete with forest and brook beauties. It is called Tempé, and will one day be a Mecca to the nature-loving tourists.

" *Idlewild*," the residence of the poet N. P. Willis, is hidden from view now, only by the front of Butter Hill; and were it not for the forest of verdure around it we might descry "Undercliff," the home of George P. Morris, near the village of Cold Spring, across the river on the east.

Cold Spring and " Undercliff."—Cold Spring is one of the most picturesque of the villages of the Hudson, whether seen from the water or from the hills behind, or in detail amidst its little streets and villa homes. It is built upon a steep ascent, and behind it is the massive granite crown of Bull Hill. This noble mountain overshadows the beautiful terrace upon which the poet Morris has lived in the rural seclusion of "Undercliff" for many years. It is scarcely possible to find a spot of sweeter natural attractions than the site of Undercliff, looking over the pretty village to the castellated hills of West Point, across the blue Hudson to old Cronest, or northward beyond the Newburgh Bay, to the far away ranges of the Catskills.

The West Point Iron Foundry, which is located here, supports much of the population and business of the village.

Two miles below are the Indian Falls, a romantic cascade, on the Indian Brook, a wild rocky stream which enters the river hereabouts.

The Beverly House, memorable for its associations with the history of the treason of Arnold, is a few miles below. See previous pages for further mention of this locality. The population of Cold Spring is 1,200. Its distance from New York, 54 miles.

Beyond Cold Spring, and still on the east bank of the river, the Highland range is continued in the jagged precipices of the Break Neck, and Beacon Hills, in height, respectively, 1,187 and 1,685 feet. These mountains are among the most commanding features of the river scenery. As we leave them to the south we approach Pollopel's Island, and enter the wide Newburgh Bay, with the villages of Cornwall, New Windsor and Newburgh upon our left, and Fishkill on our right, all imposingly displayed from the water.

Cornwall Landing, on the west bank, comes first to our reach. It is a rugged and picturesque little place. On the lofty Highland Terrace back, is Canterbury, a quiet village, much in favor as a summer residence by the seekers of repose and rural pleasure, rather than of fashionable display and distraction.

" Idlewild," Mr. Willis's romantic home, on a lofty plateau above and north of the village, is the chief object of interest. A wonderful ravine, full of the most delightful cascades, with its neighborhood of hill side, rock and forest, occupies one part of the domain, and a fertile terrace sweep, upon which his cottage stands, fills the rest. In its multiplicity of charms, it is a retreat which any poet might be content to enjoy.

There is an extensive paper manufactory, under the conduct of Mr. Carson, just back of Idlewild, in the out-of-the-way little village of Moodna. The Moodna Creek, a romantic stream, comes into the river at the northern point of Idlewild.

New Windsor, between Idlewild and Newburgh, and once the rival of the latter, is a straggling hamlet, of no special present attraction; though it has some old historical memories of interest. The chief camp ground of the Revolutionary army, during the operations on the Hudson, lies back of it, with memories and scenes yet remaining,

The Highlands—Cornwall Landing.

of the residence of Greene and Knox, and other distinguished generals of the period; of the site of the memorable old building which was known as the Temple, and was erected at the command of Washington for a chapel for the army; a hall for the free mason fraternity, which existed among the officers, and for general public assemblies. This structure wss baptized the "Temple of Virtue," at the time of its erection, a name which it lost even in the orgies of the dedicatory festival!

On the shore of Plum Point, the elegant promontoried estate of Philip A. Verplanck, Esq., at the mouth of the Moodna creek and the river, are preserved some curious débris of old military defences, and of buildings long before the days of the Revolution.

Washington established his headquarters at New Windsor, first on June 23d, 1779, and again in 1780. His residence, a plain Dutch house, has long since passed away.

"Cedar Lawn." Joel T. Headley, the distinguished author, possesses a charming river estate, which is called "Cedar Lawn," between the villages of New Windsor and Newburgh.

Asher B. Durand, the eminent landscape painter, at one time possessed and occupied an elegant country seat in the same neighborhood.

Newburgh, with its population of 12,000, and its social and topographical attractions, is one of the largest and most delightful towns on the Hudson. Rising as it does, rather precipitously from the water to an elevation of 300 feet, it presents a very imposing front to the voyager. The higher grounds are occupied by beautiful residences, and the luxurious villas of gentlemen retired from metropolitan life. There are a dozen churches, and half-a-dozen banks here, and nearly as many newspapers. Newburgh is the eastern terminus of a branch of the Erie Railway, connecting daily with that great thoroughfare at Chester, N. Y. It is united by steam ferry to Fishkill, on the opposite shore, and here is its station on the Hudson River, and Hartford, Providence, and Fishkill Railroads. It has large manufactories of various kinds, and an extensive trade in farm and dairy products. The home of the lamented landscape gardener and horticultural writer, A. J. Downing, was here. The village, too, is honored by the residence of H. K. Brown, the eminent sculptor.

Newburgh was the theatre of many interesting events in the war of the Revolution. It was the site of one of Washington's chief head-quarters, and the house in which he lived, is now the principal boast of the town. It occupies a bold position, overlooking the great pass of the Highlands. It was here that the Revolutionary army was finally disbanded at the close of the war, June 23d, 1783.

Hotels. The *Powelton* is an elegant summer house, picturesquely located in the upper and more rural part of the village. In the business centre is the *Orange Hotel*, a large and well-ordered establishment of old fame. Near the river landings, is the *United States*, an unexceptionable and comfortable place.

Fishkill, on the eastern shore, opposite Newburgh, is, like that village and all the region round, opulent in natural beauties, and prolific in elegant residences of retired city gentlemen. It is a small place, with a population, in 1854, of 1,600. It lies in the lap of a lovely fertile plain, which reaches far back to the base of a bold mountain range. It is, like all the neighborhood, replete with memories of Revolutionary and Ante-revolutionary interest. A portion of the Continental army was encamped here. The building occupied as barracks was the property of a Mr. Wharton, and has thence been since known as the Wharton House. It is, like most of the buildings of the period, a plain, Dutch, wooden construction. It may be found about half a mile south of the village.

Fishkill is the scene of many of the incidents in Cooper's novel of The Spy; a Tale of the Neutral Ground. Enoch Crosby, who was supposed to be the actual character represented in Mr. Cooper's tale, as Harvey Birch, was subjected to a mock trial before the Committee of Safety in the Wharton House, mentioned above.

Two miles north-east of Fishkill landing, is the Verplanck House, interesting as having once been the head-quarters of the Baron Steuben, and the place in which the famous *Society of the Cincinnati* was organized, 1783.

Fishkill is to be connected with Boston by the Fishkill, Hartford, and Providence Railway.

Low Point, three miles above Fishkill landing, is a small river hamlet.

New Hamburg comes next, near the mouth of Wappinger's Creek, and a little north is the village of *Marlborough*, with *Barnegat,* famous for its lime-kilns, two miles yet beyond.

Poughkeepsie is 75 miles from New York, and thus the half way station on the river railroad. It is a pleasant city, and the largest place between New York and Albany. Its population is some 15,000. It contains about sixteen churches, four banks, and three or four newspapers. It has a variety of manufactories; and the rich agricultural region behind it makes it the depôt of a busy trade.

College Hill, the site of the Collegiate Institute, half a mile north-east, is a commanding elevation, overlooking the river and the region around.

Poughkeepsie was founded by the Dutch more than 150 years ago. It is symmetrically built, chiefly upon an elevated plain, half a mile east of the river. It has no historic associations of especial interest. Professor Morse, the inventor of the electric telegraph, and Benson J. Lossing, the historian, reside here.

New Paltz Landing is on the opposite side of the river, west.

Hyde Park, and **"Placentia."**— Hyde Park, 80 miles above New York, is a quiet little village on the east side, in the midst of a country of great fertility, and thronged with wealthy homesteads and sumptuous villas. Near the village, on the north is "Placentia," famous as the home of the late veteran author, Jas. K. Paulding. Here that distinguished pioneer in American letters reached a kindly age, his time divided between his books and his fields. Placentia commands a magnificent view of the river windings, far above, even to the peaks of the distant Catskills.

Staatsburg is upon the railway, a few miles above.

Rondout and Kingston lie on the western side, the former on the Rondout Creek, one mile from the Hudson, and the latter on an elevated plain, three miles distant from the river. At Rondout is the terminus of the Delaware and Hudson Canal, through which large supplies of coal are brought to market. The Rondout Creek is a singularly picturesque stream, in all its course from the mountains, westward.

Kingston is a thriving and pleasant place. Its population in 1855,was nearly 5,000, and that of the township, 13,000. It was settled by the Dutch (1663) about the time of the settlement of Albany and New York. In the times of the Revolution it was burned by the British (1777). The first constitution of New York was framed and adopted in a house still standing here.

Kingston was the birth-place of Vanderlyn, the eminent painter. He died here in 1853.

Rhinebeck is on the railway, opposite Kingston, and is connected with that village by a ferry.

In our voyage up the Hudson, we have now, as we have had for some miles back, new and magnificent features in the landscape. Far away on the west, lie the bold ranges of the Shawangunk and the Catskill Mountains, forming fresh and charming pictures at every step of our progress.

Saugerties and Tivoli, the one on the west, and the other on the east bank of the river, now attract our attention. Saugerties is a picturesque and prosperous village, at the debouchure of the beautiful waters of Esopus Creek.

Passing *Malden* on the left, and *Germantown* on the right, we come to *Oakhill*, the station on the railway for the opposite town of

Catskill, at the mouth of the Catskill Creek, on the west bank of the

Hudson. In its pictorial attractions, this is one of the most interesting points of our present route. The village, which is a pleasant and thriving one, rises from the margin of the creek, to an elevated site on the north, where it is dissipated in many beautiful country villas, overlooking the river on the east, and the valley and mountains on the west.

Among these homes is that of the family of the painter, Thomas Cole. This great artist was buried in the village cemetery here. His studio, seen from the water, is still preserved in all its arrangements, as it was when he last occupied it.

Catskill is chiefly interesting to the tourist as the point of détour towards the wonders of the mountain ranges, which lie over the intervening valley, 10 miles westward. See Tour to the Catskill Mountains.

Hudson. In the voyage above and below Hudson, there are displayed some of the finest passages of the river scenery. With a varied shore on the east, and the Catskill peaks and ridges on the west, the tourist will scarcely regret that he has left even the Highlands behind him. Passing Mount Merino, about four miles above Catskill, the city of Hudson, lying upon the water and upon a high terrace, spreading away to higher lands on the east, comes imposingly into view. It is one of the most important river towns commercially, and one of the most attractive topographically and pictorially. The main street, which lies through the heart of the city, from east to west, terminates at the river extremity in a pleasant little park called Promenade Hill, on a bold promontory, rising abruptly 60 feet above the water; while the other terminus climbs to the fuot of Prospect Hill, an elevation of 200 feet. From these lofty heights the views of the Catskills, of the far-spreading river, and of the beautiful city itself are incomparable. There are nearly a dozen churches, some of them elegant structures, in Hudson ; a fine court-house of marble, and other public edifices, among them a famous Lunatic

Asylum. It has various educational establishments, and newspapers and other publications to the number of half a dozen. Hudson is a depôt of large business, and at one time it had an extensive India and whaling trade. It is at the head of ship navigation on the river. There are also large manufacturing interests here, maintaining upwards of seventy establishments of various kinds. It is the chief terminus of the Hudson and Boston Railway, extending eastward 34 miles to West Stockbridge, Mass., and uniting with the trains from Albany to Boston, and with other routes.

Passengers for the Shaker Village at New Lebanon, 36 miles from Hudson, take the Hudson and Boston cars to within seven miles of the Springs, which are much sought in summer time.

Columbia Springs, five miles distant, is a summer resort of great value to invalids, and of interest to all. The Claverack Falls, some eight miles off, should not be overlooked by the visitor.

Hotels. The Worth House on Main street, in the centre of the city, is an elegant establishment. There is also a good hotel near the railway station and steamboat landing. The population of Hudson in 1850, was 6,209.

Athens is a little village, with a population of 1,400, directly opposite Hudson, and connected with it by a steam ferry.

Stockport, Coxsackie and *Stuyvesant* come now in succession along the east side of the river. These are bustling and thriving little places.

Kinderhook Landing, and **"Lindenwold."** The village of Kinderhook, about five miles east of the landing, on the east side of the river, is the birthplace of Martin Van Buren, Ex-President of the United States. His present residence is upon his estate of "Lindenwold," two miles south of the village.

New Baltimore and Coeymans are now passed on the left, and Schodack and Castleton on the right, after which we yet journey some eight miles, and then reach East Albany, where we may continue on to Troy, or cross the river by ferry to the end of our present route at the city of

Albany. We are now at the capital of the Empire State, after our voyage of 145 miles (by railway, 144), from the city of New York. For the continuation elsewhere of our travels from this point to Boston, Canada, Saratoga Springs, Niagara Falls, and the Great West—for railways in all directions meet here—the tourist is referred to our Index of routes and places.

Albany was founded by the Dutch, first in a trading post on Castle Island, directly below the site of the present city, in 1614. Fort Orange was built where the town now stands, in 1623; and, next to Jamestown in Virginia, was the earliest European settlement in the original thirteen States. It was known as Beaver Wyck, and as Williamstadt, before it received its present name in honor of James, Duke of York and Albany, afterwards James the Second, at the period when it fell into British possession, 1664. The population in 1855, was about 60,000.

It has a large commerce, from its position at the head of sloop navigation and tide water upon the Hudson, as the *entrepôt* of the great Erie Canal from the west, and the Champlain Canal from the north, and as the centre to which many routes and lines of travel converge. The boats of the canal are received in a grand basin constructed in the river, with the help of a pier 80 feet wide, and 4,800 feet long.

Albany, seen from some points upon the river, makes a very effective appearance, the ground rising westward from the low flats on the shore, to an elevation of some 220 feet, in the range of a mile westward. State street ascends in a steep grade from the water to the height crowned by the State Capitol.

Hotels. The Delevan House and others.

Among the public buildings are the Capitol, the State Hall, the City Hall, the Hospital, the Penitentiary (a model prison), the Alms House, and more than

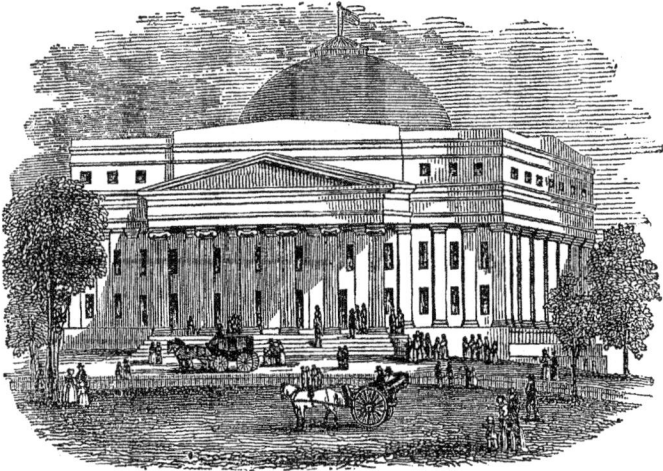

The State Hall Albany.

40 church edifices. Of the latter the new Cathedral is a noble structure.

The Dudley Observatory, founded by the munificence of Mrs. Blandina Dudley, was erected at a cost of $25,000, and has been further endowed to the amount of $100,000. The University of Albany was incorporated in 1852. The Law Department is now one of the best in the Union.

The Medical College, which was founded 1839, is a prosperous establishment. The State Normal School was organized successfully in 1844, " for the education and practice of teachers of common schools, in the science of education, and the art of teaching." The Albany Institute, for scientific advancement, has a library of 5,000 volumes. The Young Men's Association has a collection of 8,000 volumes ; the Apprentices' Library, 3,000 ; and the State Library (accessible to public use), has 46,000 volumes.

The edifice on State street where are deposited the public collections in natural history, and in geology, and in agriculture, is most interesting. The Orphan Asylum, and other benevolent establishments of the city, are well worth the consideration of the tourist.

The distinguished sculptor, E. D. Palmer, resides here. His studio is a place of especial attraction.

Troy is a large and beautiful city of over 45,000 inhabitants. It is upon both banks of the Hudson, at the mouth of the Poestenkill Creek. It is built upon an alluvial plain, overlooked on the east side, by the classic heights of Mount Ida, and on the north by the barren cliffs of Mount Olympus, 200 feet high. These elevated points command superb views of the city and its charming vicinage, and of the great waters of the Hudson. Troy lies along the river for the length of three miles, and drops back a mile from east to west. Troy is a busy city, with its manufacturing industry, and as a great entrepôt of railway travel from and to all points. It boasts many fine churches and public buildings, and many admirable private mansions and cottages. Here is the well-known Female Seminary, established by Mrs. Emma Willard, in 1821. It is the seat, too, of the Troy Polytechnic Institute.

The chief hotels of Troy are the American Hotel, Mansion House, Troy House, Temperance House, Northern Hotel, Washington Hall, Union Hall, and the St. Charles. The cars leave Troy and Greenbush every hour during the day and evening. Steamboats and stages also run between Albany and Troy. Railway trains extend to all points. See Index.

West Troy, a suburb of Troy, on the opposite side of the river, is a rapidly growing place. The inhabitants are employed principally in manufactures. A fine Macadamized road leads from West Troy to Albany, a distance of six miles.

At **Gibbonsville** is a United States Arsenal, where is kept a large and constant supply of small arms, and the various munitions of war. This is one of the most important of the national depots, and is worthy the attention of the traveller.

NEW YORK TO LAKE ERIE,

By the New York and Erie Railroad.

This great route claims especial admiration for the grandeur of the enterprise which conceived and executed it, for the vast contribution it has made to the facilities of travel, and for the multiplied and varied landscape beauties which it has made so readily and pleasantly accessible. Its entire length, from New York to Dunkirk, on Lake Erie, is 460 miles (including the Piermont and the Newburgh branch, it is 497 miles), in which it traverses the southern portion of the Empire State in its entire extent from east to west, passing through countless towns and villages, over many rivers; through rugged mountain passes now, and anon amidst broad and fertile valleys and plains. In addition, it has many branches, connecting its stations with other routes in all directions, and opening yet new stores of pictorial pleasures.

The road was first commenced in 1836. The first portion (46 miles, from Piermont to Goshen) was put in operation September 23d, 1841; and, on the 15th of May, 1851, the entire line to Lake Erie was opened amid great rejoicings and festivals, in which the President of the United States and other distinguished guests of the company assisted.

Some idea of the extent of this noble route may be gathered from the fact, that, in 1854, it employed about 200 locomotives, nearly 3,000 cars, 4,000 employés (682 of which are engaged in repairing engines and cars). The cost of the road and equipments, up to 1854 (including the Newburgh branch), was nearly $34,500,000. The earnings for the year 1856 were $6,349,050 15, and the expenses for the same period were $5,002,754 45.

An interesting feature of this road, and one of great convenience to the Company and security to the traveller, is its own telegraph, which runs by the side of the road through its whole extent, and has its operator in nearly every station-house. This telegraph has a double wire the entire length of the road; enabling the Company to transact the public as well as their own private business. Daily trains leave for the West on this route, from the foot of Duane street, morning, noon, and night.

STATIONS.

NEW YORK, Jersey City, Bergen, 2 miles; Boiling Spring, 9; Passaic Bridge, 11; Huyler's 12; PATERSON, 16; Godwinville, 21; Hohokus, 23; Allendale, 25; Ramsey's, 27; Suffern's, 31; Ramapo, 33; Sloatsburg, 35; Southfields, 41; Greenwood, 44; TURNER'S, 47; Monroe, 49; Oxford, 52; Chester, 55; Goshen, 59; Hampton, 63; Middletown, 66; Howell's 70; Otisville, 75; PORT JERVIS, 88; Shohola, 106; LACKAWAXEN, 110; Mast Hope, 116; NARROWSBURG, 122; Cochecton, 130; Callicoon, 135; Hankins, 142; Basket, 146, Lordville, 153; Stockport, 159; Hancock, 163; Hale's Eddy, 171; DEPOSIT, 176; SUSQUEHANNA, 192; GREAT BEND, 200; Kirkwood, 206; Conklin, 210; BINGHAMTON, 214; Hooper, 220; Union, 223; Campville, 228; OWEGO, 236; Tioga, 242; Smithboro, 246; Barton, 248; Waverley, 255; Chemung, 260; Wellsburg, 266; ELMIRA, 273; Junction, 277; Big Flats, 283; CORNING, 291; Painted Post, 292; Addison, 301; Rathboneville, 306; Cameron, 314; Adrian, 322; Canisteo, 327; HORNELLS-

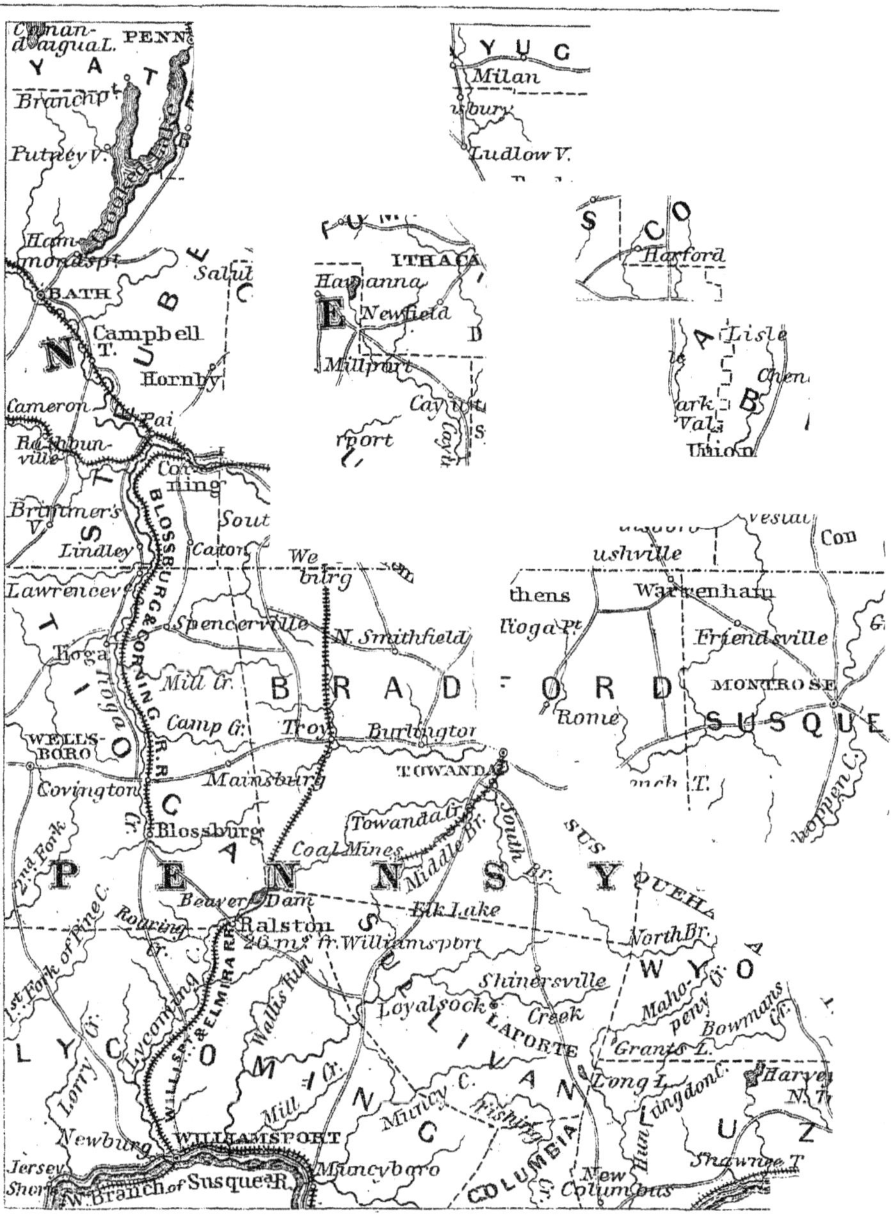

Canandaigua L.
PENN
Y A T
Branchpt
Putney V.
Hammondspt
BATH
Campbell T.
Hornby
Cameron
Rathbunville
Corning
BLOSSBURG & CORNING R.R.
Lindley
Caton
Lawrenceville
Spencerville
N. Smithfield
Tioga
Mill Cr.
Camp Cr.
Troy
Burlington
B R A D F O R D
WELLS-BORO
Covington
Mainsburg
TOWANDA
Towanda Cr.
Blossburg
Coal Mines
Middle Br.
South Br.
P E N N S Y
Beaver Dam
Ralston
26 m. fr. Williamsport
Elk Lake
Roaring Cr.
Lycoming C.
WILLIAMSPT & ELMIRA R.R.
Wallis Run
Shinersville
Loyalsock Creek
LAPORTE
Grants L.
Long L.
Bowmans Cr.
Mill Cr.
Muncy C.
Fishing
Lorry Cr.
Newburg
WILLIAMSPORT
Jersey Shore
W. Branch of Susquea R.
Muncyboro
COLUMBIA
New Columbus
Shawnee T.
Harvey
ITHACA
Havanna
Newfield
Millport
Milan
Ludlow V.
Harford
Lisle
Union
Warrenham
Friendsville
MONTROSE
Rome
Tioga Pt.
SUSQUE

VILLE, 331; Almond, 336; Alfred, 340; Andover, 349; Genesee, 358; Scio, 361; Phillipsville, 365; Belvidere, 369; Friendship, 373; Cuba, 382; Hinsdale, 389; OLEAN, 395; Allegany, 398; Great Valley, 411; Little Valley, 420; Cattaraugus, 428; Dayton, 437; Perrysburg, 440; Smith's Mills, 447; Forestville, 451; Sheriden, 455; DUNKIRK, 460.

To Suffern's (31 *miles*) *via New Jersey, and via Piermont.* The first 31 miles of the Erie Route, that through the State of New Jersey, from Jersey City, opposite New York, to "Suffern's," consists of parts of three different railways, though used of late years for all the general passenger travel of the Erie road, and with its own broad track and cars. The original line of the road is from Suffern's eastward, 18 miles, to Piermont, and thence 24 miles down the Hudson river. This route is now employed only for freight, and for local travel. It leads through a rude but not uninteresting country, with here and there a fine landscape or an agreeable village.

We pass now, without halt, through the New Jersey towns—Patterson, with its "Falls of the Passaic" among them, and begin our mention of places and scenes of interest on the Erie route, at Suffern's station, where the original Piermont and the present Jersey City lines meet. The Ramapo Valley commences at this point, and, in its wild mountain passes, we find the first scenes of especial remark in our journey. Fine hill farms surround us here, and on all our way through the region of the Ramapo for 18 miles, by *Sloatsburg, Southfields, Greenwood,* and *Turner's* to *Monroe.* The chief attraction of the Ramapo Gap is the Torn Mountain, variedly seen, on the right, near the entrance of the valley and about the Ramapo station. This is historical ground, sacred with memories of the movements of the Revolutionary army, when it was driven back into New York from the Hudson. Washington often ascended to the summit of the Torn to overlook the movements of the

British. On one such occasion, anecdote says that he lost his watch in a crevice of a rock, of which credulity afterwards heard the ticking in the percolations of unseen waters. Very near the railway at Suffern's the debris of old intrenchments are still visible; and marks of the camp fires of our French allies of the period may be traced in the woods opposite. Near by is an old farm house, once occupied by the commander-in-chief. The Ramapo is a great iron ore and iron manufacture region; and it was here that the great chain which was stretched across the Hudson to check the advance of the English ships, was forged, at the spot once called the Augusta Iron Works, and now a poetical ruin by a charming cascade with overhanging bluff, seen close by the road, on the right, after passing Sloatsburg. The Ramapo Brook winds attractively through the valley, and beautiful lakelets are found upon the hill tops. There are two such elevated ponds near Sloatsburg. At Sloatsburg passengers for the summer resort of Greenwood Lake, 12 miles off, take stage tri-weekly. See Greenwood Lake.

From *Monroe* onward through *Oxford, Chester, Goshen, Hampton, Middletown, Howell's,* and *Otisville* to *Port Jervis* (or Delaware) we are in the great dairy region of Orange County, New York, which sends a train of cars laden only with milk daily to the New York market. A very charming view is seen south from the station at Oxford, led by the cone of the Sugar Loaf, the chief hill feature of the vicinage. At Chester, the branch road from Newburgh, on the Hudson river, 19 miles long, comes in. From this point, as well as from Sloatsburg, passengers for Greenwood Lake (eight miles) take stage. At Howell's, 70 miles from New York, the country gives promise of the picturesque displays to be seen through all the way onward to Port Jervis. Approaching Otisville, the eye is won by the bold flanks of the Shawangunk Mountain, the passage of which great barrier (once deemed almost insurmountable) is a miracle of engineering skill. A mile beyond Otisville, after traversing an ascending grade of 40 feet to the mile, the road runs through a rock cutting, 50 feet deep and 2,500 feet long. This passed, the summit of the ascent is reached, and thence we go down the mountain side many sloping miles to the valley beneath. The scenery along the mountain slope is grand and picturesque, and the effect is not lessened by the bold features of the landscape all around—the rugged front of the Shawangunk, stepping, like a colossal ghost, into the scene for one instant, and the eye anon resting upon a vast reach of untamed wilderness. In the descent of the mountain the embankment is securely supported by a wall 30 feet in height and 1,000 feet long. The way onward grows momently in interest, until it opens upon a glimpse, away over the valley of the mountain spur, called the Cuddleback; and, at its base, the glittering water seen now for the first time, of the Delaware and Hudson Canal, whose *débouchure* we have looked upon at Kingston, in our voyage up the Hudson river. Eight miles beyond Otisville we are imprisoned in a deep earthy cut for nearly a mile, admirably preparing us for the brilliant surprise which awaits us. The dark passage made, and yet another bold dash through rocky cliffs, and there lies suddenly spread before us, upon our right, the rich and lovely valley and waters of the Neversink. Beyond sweeps a chain of blue hills, and at their feet, terraced high, there gleam the roofs and spires of the village of Port Jervis; while onward, to the south, our eye first beholds the floods of the Delaware, which is to be so great a source of delight in all our journey hence, for nearly 90 long miles, to Deposit.

Port Jervis, or Delaware, as the station was called, is the terminus of the eastern division, one of four great subsections into which the road is measured. It is the point at which the tourist who can spend several days in viewing the route, should make his first night's halt. The vicinage is replete with pictorial delights, and with ways and means for rural sports and pleasures. Charms of climate and of scenery, with the additional considerations of a pretty village and a most excellent hotel (the Delaware House at the station), have made Port Jervis a place of great and continuous summer resort and tarry.

There is a stage route hence, 6 miles, to the neighboring "Falls of the Sawkill." This stream, after flowing sluggishly for some miles through level table-land, is here precipitated over two perpendicular ledges of slate-rock—the first of about 20 feet, and the second about 60 feet—into a wild gorge. The brook still continues, dashing and foaming on for a quarter of a mile, over smaller precipices, and through chasms scarcely wide enough for the visitor to pass. The beetling cliffs that form the sides of the gorge are surmounted and shaded by cedars and hemlocks, that lend a peculiarly sombre air to the scenery. The sojourner here must not omit a tramp to the top of Point Peter, overlooking the village, and all the wonders for miles around.

We now continue the transit of the second grand division of the road, which carries us onward, 104 miles further, to *Susquehanna*, and from New York, all told, 192 miles. The canal keeps us company, nearer or more remote, for some miles, and by and by we cross the Delaware on a fine bridge of 800 feet, built at a cost of $75,000. The river, from this point, is seen, both above and below, to great advantage. Here we leave Orange County and New York for a little incursion into the Keystone State, for which privilege the company pays Pennsylvania ten thousand dollars per year.

The canal, and its pictures and incidents, are still the most agreeable features of our way, though at Point Eddy we open into one of the wide basins so striking in the scenery of the Delaware.

Near *Shohola* (106 miles from New York), we are among some of the greatest engineering successes of the Erie route, and some of its chief pictorial charms. Here the road lies on the

The Delaware River, near Hancock, N. Y. & Erie Railway.

mountain side, several feet above the river, along a mighty gallery, supported by grand natural abutments of jagged rock. It is a pleasant scene to watch the flight of the train upon the crest of this rocky and secure precipice; and the impressiveness of the sight is deepened by its contrast with the peaceful repose of the smiling meadow slopes, on the opposite side of the river below. Upon three miles alone of this Shohola section of the road, no less than three hundred thousand dollars were expended.

At Lackawaxen (111 miles from New York), there is a charming picture of the village, and of the Delaware bridged by the railway, and by the grand aqueduct for the passage of the canal, supported by an iron wire suspension bridge.

We pass on now by Mast Hope to Narrowsburg.

Narrowsburg (122 miles from New York and 338 from Dunkirk), is a pleasant place for quiet summer rest and rural pastime, with its inviting hotel comforts, and its piscatory and field recreations and sports.

Beyond Narrowsburg, for some miles, the traveller may turn to his newspaper or book for occupation a while, so little of interest does the scene, without, present, with the exception, now and then, of a pleasing bit of pastoral region. Some compensation may be found in recalling the stirring incidents of Cooper's novel of the "Last of the Mohicans," of which this ground was the theatre.

At Callicoon, a brook full of wild and beautiful passages and of bright trout, comes down to the Delaware.

As we approach Hancock, once called Chehocton, we come near the charming picture of the meeting of the two branches of the Delaware, seen on our left.

Hancock is one of the most important places on this division of our route, and in every way a pleasant spot for sojourn.

At *Deposit*, 13 miles beyond Hancock, and 176 from New York, we bid goodbye to the Delaware, which we have followed so long; refresh ourselves at the excellent café, and prepare for the ascent of a heavy grade over the high mountain ridge which separates it from the lovely waters of the Susquehanna. We go up 58 feet per mile to an elevation 365 feet above Deposit. The way is wild and desolate, covered with the jagged *debris* left in the strong battle with the mountain fastnesses. The grand pass of the Summit reached, we descend again by a grade of 60 feet, into that most beautiful region of the Erie road, the Valley of the Susquehanna.

For a little while, as we go down, there seems no promise of the wonders which are awaiting us, but they come suddenly, and before we are aware, we are traversing the famous

Cascade Bridge, a solitary arch, 250 feet wide, sprung over a dark ravine of 184 feet in depth.

No adequate idea of the bold spirit and beauty of the scene can be had from the cars, and especially in the rapid transit often passed before the traveller is aware of its approach. It should be viewed leisurely from the bottom of the deep glen, and from all sides, to be realized aright. To see it thus, a half a day's halt should be made at the next station, to which we shall soon come.

The Cascade Bridge crossed, the view opens almost immediately, at the right—deep down upon the winding Susquehanna, reaching afar off amidst a valley and hill-picture of delicious quality, a fitting prelude to the sweet river scenes we are henceforth to delight in. This first grateful glimpse of the brave Susquehanna is justly esteemed as one of the finest points on the varied scenery of the Erie Railway route. It may be looked at more leisurely and more lovingly by him who tarries to explore the Cascade Bridge hard by, and the valley of the Starrucca, with its grand viaduct, which we are now rapidly approaching.

The Starrucca Viaduct (190 miles from New York and 250 from Dunkirk), is one of the chief art-glories of our present route—perhaps the chiefest. The giant structure is made of stone from the ravine, two miles above, crossed by the fairy Cascade Bridge. It is 1,200 feet in length and 110 feet high, and has 18 grand arches, each 50 feet span. The cost was $320,000. The landscape is of exceeding beauty, whether seen from the viaduct or from any one of many points, near or afar off, below. From the vicinity of Susquehanna, the next station,

Cascade Bridge, Erie Railway.

the viaduct itself makes a most effective feature in the valley views.

A little way beyond, and just before we reach the Susquehanna station, we cross a fine trestle bridge, 450 feet long, over the Cannewacta Creek, at Lanesborough. We are now fairly upon the Susquehanna, not in the distance, but near its very marge, and, anon, we reach the end of the second grand division of our route, and enter the busy depot of Susquehanna, from New York 192 miles, and from Dunkirk 267.

At **Susquehanna** we are passing beyond the wild scenery on our route, and in a few miles further we shall fall in with and follow, for many miles, through broad valley tracks, coursed by the great winding river—a country which we shall find replete with interest, and very often of marked natural beauty, however unlike the scenes upon which we have looked in our transit of the wild hills and forest region of the Delaware.

The Susquehanna station is one of the busiest points on our route, being the place where divisions meet—where the great massive engines, or *pushers*, which are used to *push* the heavy trains hence to the top of the grand hill "Summit," are housed, and where the workshops for the repairs of disabled locomotives and cars are located. 200 hands are employed here by the company. Indeed, the place is all railroad, from which it was born and from which it has grown. If the hotels at this station are too noisy for the tarrying stranger, we may go a mile backwards to Lanesborough, and from thence review the scenes of the Starrucca and of the Cascade Bridges, with many other points of pictorial attraction.

Just beyond the Susquehanna depot we cross to the right bank of the river, and, after two more miles ride, yet amidst mountain ridges, we reach

Great Bend, 200 miles from New York, and 259 from Dunkirk. The village of this name lies close by, in the State of Pennsylvania, at the base of a bold cone-shaped hill.

At Great Bend there comes in to the Erie Road the Delaware, Lackawanna, and Western Railway, leading nearly south into Pennsylvania, through the coal regions of Scranton, the neighborhood of the valley of Wyoming, the Water Gap of the Delaware, and ending upon that river, five miles yet below ; here it is connected by other railway routes with New York, Philadelphia, etc. See Index.

Leaving Great Bend we enter upon the more cultivated landscape of which we lately spoke, and approach villages and towns of greater extent and elegance.

Near Kirkwood, the next station, six miles from Great Bend, there stands an old wooden tenement, which may attract the traveller's notice as the birthplace of the Mormon prophet, Joe Smith.

Binghamton, 216 miles from New York, is, with its population of ten or eleven thousand people, one of the most important places on the Erie route, and indeed in southern New York. It is a beautiful town, situated upon a wide plain, in an angle made by the meeting of the Susquehanna and the Chenango rivers.

Binghamton was settled in 1804, by Mr. Bingham, an English gentleman, whose daughters married the brothers Henry and Alexander Baring, the famous London bankers. One of those gentlemen was afterwards created Lord Ashburton.

The Chenango Canal, extending along the Chenango river, connects Binghamton with Utica, N. Y., 95 miles distant : and it is also the southern terminus of the Syracuse and Binghamton Railroad, 80 miles long. See Index. Passing on by the stations of Hooper, Union, and Campville, we come to

Owego, another large and handsome town, almost rivaling Binghamton in beauty and importance. Owego is surrounded by a landscape, not of bold but of very beautiful features. Many noble panoramas are to be seen from the hill-tops around, overlooking the village and the great valley. The Owego

Creek, which enters the Susquehanna here, is a charming stream. Just before its meeting with the greater waters, it passes through the meadow, and at the base of the hill slopes of "Glenmary," once the home of N. P. Willis, and now one of the Meccas of the vicinage, to which all visitors are won by the charms and spells the genius of the poet has cast about it. It was here that Mr. Willis wrote his famous "Letters from under a Bridge."

The Cayuga and Susquehanna Railroad diverges here, some 30 miles, to Ithaca, on Cayuga Lake. See Index.

The Owaga House, in the heart of the town and on the banks of the Susquehanna, is a large and elegant summer hotel.

Elmira.—Passing the half dozen intermediate stations, we jump now, 37 miles, to Elmira, 273 miles from New York. This beautiful town is a peer of Binghamton and Owego, with the same charming valley nest and the same environing hill-ridges.

The Newton Creek and the Chemung River, near the junction of which waters Elmira is built, lend a world of picturesque beauty to the vicinage.

The Elmira, Canandaigua, and Niagara Falls Railway diverges here, and connects the village with the Canada lines. This road is one of the pleasantest from New York to the Falls of Niagara.

The Williamsport and Elmira Railroad conducts hence into Pennsylvania, and unites with other lines for Philadelphia. The Chemung Canal also connects Elmira with Seneca Lake, 20 miles distant. It is a delightful excursion from the village to Geneva and other places on the Seneca Lake, by the railway travel. See Index for routes from Elmira.

Five miles beyond Elmira, our route is over the Chemung River, bringing us to "Junction," the starting point of the Chemung Railroad for Jefferson and Niagara.

Corning, 290 miles from New York, is an important point on the Chemung River. The feeder of the Chemung Canal extends hither from Elmira. It is the depôt of the Corning and Blossburg Railroad, hence 40 miles to coal beds of Pennsylvania. At Corning there terminates also the Buffalo, Corning, and New York Railroad, 94 miles, via Avon (Springs), and Batavia to Rochester, on the great routes west from Albany. See Index.

Hornellsville. Passing half a dozen stations, we now reach Hornellsville, where passengers for Buffalo, Niagara, &c., follow the Branch Road north, for 91 miles. (See Buffalo Division.) At this point we enter upon the fourth and last division of the Erie route, being now 331 miles from New York, and having 128 miles yet to travel to Dunkirk. The country through the rest of our way is comparatively new, and no important towns have yet grown up within it. Pictorially this division is the least attractive of the whole route, though beautiful scenes occur still, at intervals, all along. Beyond Hornellsville, we enter the valley of the Caniacadea, a fine mountain passage, filled with the merry waters of the Caniacadea Creek. *Almond* and *Alfred* lie upon the banks of this charming stream.

Reaching Tip Top Summit (the highest grade of the Erie Road, being 1,700 feet above tide water), we commence the descent into the valley of the Genesee. The country has but few marks of human habitation to cheer its lonely and wild aspect, and for many miles onward, our way continues through a desolate forest tract, alternated only by the stations and little villages of the road. Beyond Cuba Summit, there are many brooks and glens of rugged beauty.

Passing Olean, on the Alleghany River, we come into the lands of the Indian Reservation, where we follow the wild banks of the Alleghany, between lofty hills, as wild and desolate as itself.

At Cattaraugus, 428 miles from New York, and 41 from Dunkirk, we traverse a deep valley, where the eye is relieved for a little while, with scenes of gentler aspect than the unbroken forest we have long traversed, and are to traverse still.

from ALBANY

ATOGA and
EHALL,
rt of the
& BUFFALO
UTES.

LAKE PLEASANT
L. Pleasant
Piseco
Piseco
L.
Oxbow R.
Wells
Hope
S.Hope
Fish L.
Northampton
Cranberry

WARREN

Warren

Hud

Day
Hadley
Sacandaga R.
Edinburg
Corinth

SARATO
sh House   Greent
dence   Wil
SARATOGA
SPRINGS
alway
3

WHITE H
omstocks l
CALDWELL   Fort
Ann
61
King's
bury

ansingburg
RENSSE
Patro
Poestenville
h
Berlin
bush

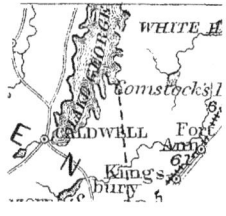

UATION
the
& BUFFALC
TES
ACUSE.
Ct.

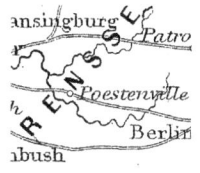

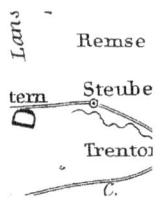

Lans
Remse
tern   Steube
D
Trento
C.

CONTINUATION
of the
ANY & BUFFALO
ROUTES
ROCHESTER.

LAKE ONTARI

Lit Sodus
East Bay
Port B
Sodus b

Sodus P.

mson
Socus
arion
A   Y
S Sor
V.
adr.

Marcellus
Falls
Marieta
Otisco
L.
L

dana

Skaneateles Lake

logsville

L. T.

9

Three miles beyond Perrysburg we catch our first peep at the great Erie waters, towards which we are now rapidly speeding. Yet a few miles, and we are out of the dreary woods, coursing again through the more habitable lands which lie upon the lakes.

Reaching **Dunkirk** at last, we may pursue our journey westward by any one of many routes by land and by water—on the blue waves, or still upon the rapid rail. We shall follow all these routes in other pages—the steamers to Cleveland or Detroit, or the lake shore road to Cleveland. Thence by railway to Columbus, and Cincinnati, southward, or to Toledo, westward. From Toledo onward by the Michigan and Northern Indiana Railway to Chicago, and thence again by the Rock Island Road to the Mississippi, or by the Illinois Central route to St. Louis, and by other ways still onward, to the far west.

NEW YORK TO BUFFALO AND NIAGARA FALLS.

VIA BUFFALO BRANCH OF NEW YORK AND ERIE RAILROAD.

Follow the main trunk of the New York and Erie Road, from Jersey City, 331 miles to Hornellsville, N. Y.

Trains continue on immediately to Buffalo, on their arrival at Hornellsville, by the Branch Route, formerly the New York City and Buffalo Railroad.

STATIONS.

HORNELLSVILLE ; Burns, 9 ; Whitney's, 13 ; Swainville, 17 ; Nunda, 24 ; Hunt's Hollow, 26 ; PORTAGE, 30 ; Castile, 34 ; Gainesville, 37 ; Warsaw, 44 ; Middlebury, 49 ; Linden, 53 ; ATTICA, 60 ; Darien, 64 ; Alden, 71 ; Town Line, 76 ; Lancaster, 81 ; BUFFALO, 91.

The road follows by the side of the Dunkirk track through the village, and then bends northward. For nearly 30 miles, along very elevated ground, there is but little to interest the tourist, until he comes in sight of the village of Portage, lying in a deep valley to the northwest.

Portage is deservedly a Mecca to the lover of the picturesque, abounding, as it does, in the wildest wonders of moun-

tain gorge and cataract. The Genesee River steals and tumbles through the lawns and ravines of this region in a very wonderful way. At Portage, it enters a grand rocky defile, and in passing, falls in many a superb cascade. Near the station, this gorge is crossed by the railroad, upon a bridge of great magnitude and remarkable construction. From below, it rises upon the view like story upon story of solid and symmetrical scaffolding, to a height of 234 feet; its length is 800 feet. Beneath its huge masses of timber, foams the river, and by its angry side are the placid waters of the Genesee Valley Canal.

To see the wonders of Portage aright, one must tarry for days in the village, or better yet, at the hotel near the station house.

The Genesee makes a bold descent of 40 feet (seen from the cars), as it rushes beneath the great bridge, onward to yet deeper beds. A quarter of a mile northward is the second cataract of 80 feet; huge high cliffs soar yet far above it. To see the scene properly, the visitor will cross the bridge over the Genesee above the mill, and place himself immediately in front of the fall.

Some distance beyond, a staircase conducts to the bottom of the ravine, whence you may pass in a boat, or pick your way along beneath the spray of the tumbling floods. The walls of this gorge are of slate stone; they rise to a height of more than 300 feet, and in the many and sudden turnings of the way, offer a grateful succession of noble pictures.

A mile and a half still down the glen, and we reach the third, and, perhaps, the grandest of the cascades; placed as it is in an exceedingly deep and narrow passage of the ravine. This leap is 60 feet.

The canal far up above the descending bed of the Genesee in this vicinage, is a most telling feature in the landscape—a strain of gentleness in the wild anthem of the rugged ravine.

We leave the traveller here to pur-

The Catskill Mountain House.

sue the rest of his way, 61 miles, to Buffalo, and to go thence to Niagara or elsewhere, as he may find directions in other parts of our Hand Book.

NEW YORK TO THE CATSKILL MOUNTAINS.

We can commend to the traveller no pleasanter or more profitable summer excursion for a day, or a month, or even a season, than a visit to the Cats-kills—one of the grandest and most picturesque of the mountain ranges of the United States.

To reach the Catskills from New York we will follow our previous routes up the Hudson to the village of Cats-kill (111 miles), or the river railway to Oakhill station opposite, crossing thence to Catskill by ferry.

At Catskill good stages are always in waiting to convey passenges to the Mountain House, on the crest of the hills, 12 miles westward. This ride will occupy about four hours, at a cost of one to one and a quarter dollars.

The Catskills are a part of the great Appalachian chain which extends through all the eastern portion of the Union, from Canada to the Gulf of Mexico. Their chief ranges follow the course of the Hudson River, from some 20 to 30 miles, lying west of it, and separated by a valley stretch of 10 to 12 miles. These peaks lend to all the landscape of that part of the Hudson from which they are visible, its great-est charm.

The Mountain House is reached by a pleasant stage coach ride through ever-changing scenes of valley and hill. The last three miles of the journey is up the side of the mountain, made easy by a good winding way. Within a mile of his destination, the tourist halts upon the spot universally conceded to be the site of the famous 15 years' nap of Mr. Irving's myth, Rip Van Winkle.

The Mountain House stands near the brink of some bold rocky ledges, upon the summit of one of the eastern ranges, commanding all the landscape round for miles and miles away. Lifting its grand front thus, it is a curious and wonderful object, no less within its own shadow than at every point from which it may be seen. It is a massive and elegant structure of wood, with a grand facade of columns reaching the entire height of the eaves. It was originally built by the people of Catskill, at an expenditure of 20,000 dollars; but it has since been from time to time enlarged and improved, until now it possesses every reasonable, if not every possible, hotel convenience and comfort—capacious and well-furnished parlors, halls, and chambers—a luxurious table, and attentive hosts and waiters—and bathing, billiard, and bowling appointments. In the summer the house is a post-office, with daily mails.

The superb panorama of the river and valley of the Hudson, and of the New England hill ranges to the eastward, which the bold site of the Mountain House commands, will first fix the attention and admiration of the guest. Of this unrivalled sight he will never weary, so varied is it in the changing hours and atmospheres, and so imposing under every aspect. It is thought, at the dawn of the day and at the rising of the sun, when his magic beams are lifting the mystical vapor and cloud-curtain, which the night has invisibly spread over the scene, that the enchantment will reach its highest point. Luckily for the tourist who is not an enthusiast, but is contented with the simple, solid fact of a subject, like Mr. Gradgrind, these marvellous exhibitions of sun-rise effects may be comfortably seen from his warm, secure chamber-window, when the morning air is, as it often chances to be, at this mountain altitude, rather too chill and damp for comfort.

A visit to the locality, called the "North Mountain," will be a remunerative morning or afternoon's walk. It is only a mile or two through the forest, on the lofty ridge; but a guide will be desirable, for the path is more easily lost than found. At the end of the stroll he will look back upon his wilderness home over a brace of dainty little lakelets, smilingly sleeping on the mountain top; and beyond, towards the south and east, his eye will follow the windings of the Hudson far down in the sunny valley. Some stories may be told him of the fondness of the bear for this particular locality, but he need not be alarmed, for it is rarely indeed, except it be when the winter snows envelope the earth, that these gentry are about.

Another agreeable excursion will be in the opposite direction, from the house to the spot known as the "South Mountain," where, upon the brink of huge cliffs, may be seen the river and valley, and the wonderful pass of the Kauterskill, through the mountain chain westward.

The Two Lakes, which we have just overlooked from the North Mountain, make one of the leading items in the Catskill programme. They lie side by side, in gentle beauty, in the heart of the lofty plateau, upon the eastern brink of which the Mountain House is perched. They may be reached in a pleasant little walk back of the hotel. Onwards, and on the way to the Great Falls of the Kauterskill, a few minutes' stroll, indeed, is sufficient to bring us to the nearest of these twin waters, the Upper or Sylvan Lake. This is a spot for repeated and habitual visits, with its pleasures by the forest shore; in the skiff, upon the quiet and lonely flood; or, with angle in hand and trout in prospect.

The High Falls lie two miles back of the Mountain House, overleaping the western brink of the great plateau. A wagon road leads thither; and there is, besides, a footpath in the forest, by which the way is shortened one half. A good team is sent down with passengers (fare 25 cents) at least once a day from the hotel. At the very brink of the cascades there is another small but pleasant summer inn, called the Laurel House, kept by Mr. Scutt, the proprietor of the Falls. It is a wonderful sight to overlook the ravine below, and the giant crests of Round Top and High Peak—the proudest of all these hills—from the windows, or piazzas of the Laurel House, or from the platform in front, which overhangs the glen. This

Lake on the Catskills.

view enjoyed, with refreshments if you please, we commence the descent to the base of the cataracts, by many straggling flights of wooden steps. Coming to the base of the first Fall, we may steal along a narrow ledge behind the descending torrent, as one gets to Termination Rock, at Niagara. On the opposite bank parties often pic nic, the means and appliances, if duly ordered before at the Laurel House, being lowered down, upon a signal, in a basket, over the edge of the projecting platform above. The descent of the first cascade is 175 feet, and of the second, 75 feet, with many a tumble of the vexed waters afterwards in their way for a mile down the ravine into the main branch of the Kauterskill or Catskill Creek, which dashes down the great clove, of which the Mountain House stream is only an arm held at a right angle.

Fenimore Cooper, in his story of the "Pioneer," thus describes these cascades—"The water comes croaking and winding among the rocks, first, so slow that a trout might swim in it, then starting and running like any creature that wanted to make a fair spring, till it gets to where the mountain divides like the cleft foot of a deer, leaving a deep hollow for the brook to tumble into. The first pitch is nigh 200 feet, and the water looks like flakes of snow before it touches the bottom, and then gathers itself together again for a new start; and maybe flutters over 50 feet of flat rock before it falls for another 100 feet, when it jumps from shelf to shelf, first running this way and that way, striving to get out of the hollow, till it finally gets to the plain."

This branch of the Kauterskill comes from the waters of the two lakes on the plateau above; and, as the supply has to be economized in order that the cascades may look their best when they have company, the stream is dammed, and the flood is let on at proper times only. For this service, and for the use of the steps, perchance of guide also, every visitor pays a toll of 25 cents. This is a reasonable although a disagreeable bit of prose in the poem of the Catskill Falls.

We have now peeped at all the usual "sights" of the region; but there are other chapters of beauty, perhaps, yet more inviting. Let the tourist, if he be adventurous and is a true lover of nature, follow the brook down from the base of the cataracts we have just described, into the principal clove; then let him ascend the main stream for a mile over huge boulders, through rank woods, and by many cascades, which, if smaller, are still more picturesqe than those "nominated in the bond;' or, let him descend the creek, two miles, sometimes by the edge of the bed of the waters, and, when that is impracticable, by the turnpike road, which traverses the great clove or pass. At every turn and step there will be a new picture— sometimes a unique rapid or fall, sometimes a soaring mountain cliff, sometimes a rude bridge across the foaming torrent, sometimes a little hut or cottage, and, at last, as he comes out towards the valley on the east, the humble village of Palenville. This portion of the Catskills is that most preferred by artists for study, and the inns at Palenville are often occupied by them, though they offer no inviting accommodation to the ease and comfort-loving tourist.

At one time (when the hemlock was abundant on the mountain sides) this clove was a den of tanneries, and a few establishments of the kind yet linger here.

Stony Clove. Another nice excursion from the Mountain House, is a ride along the ridge five or six miles to the entrance of the Stony Clove, and thence on foot, or still in your vehicle (though the wagon road is execrable), through the wilderness of this fine pass.

High Peak, the most elevated of the Catskill summits, towering 4,000 feet towards heaven, should certainly be climbed, in order to see the region fairly. It is a long and toilsome journey, especially for ladies — six miles thither *on foot*—but we have accompanied the fairest of women through the difficulties and dangers of the way. Once we "assisted" at a night camp on the very crown of High Peak, of a party which included a dozen damsels. If they had not been brave, as they all were, they would not have deserved the glorious sunrise effects, which they saw never to be forgot, from their ambitious bivouac. Even the Mountain House on its grand perch, looked from High Peak like a pigmy in the vale.

Plauterkill Clove, is another grand pass on the hills, five miles below the Kauterskill passage. A mountain torrent, full of beauties in glen and rock and cascade, winds through it. A post road also traverses the pass. High Peak rises on the north of the Plauterkill, and the South Mountain, on which is a lovely lake, ascends on the opposite side. It is not yet a scene of much resort, being out of the very convenient reach of the Mountain House, and having no hotel attractions in its neighborhood.

The tourist here will recall with pleasure, Bryant's dainty poem of the Katterskills, from which we will borrow a few lines to end our own intimations —

"Midst greens and shades the Catterskill leaps
  From cliffs where the wood flower clings;
All summer he moistens his verdant steeps,
  With the light spray of the mountain springs;
And he shakes the woods on the mountain's side,
  When they drip with the rains of the autumn tide.

" But when in the forest bare and old,
  The blast of December calls,
He builds in the star-light, clear and cold,
  A palace of ice where his torrent falls,
With turret and arch and fretwork fair,
And pillars clear as the summer air."

The Cataracts of the Catskills in winter, when the spray is frozen into a myriad fantastic forms, all glowing like the prism as the clear cold sunlight reveals these mystical wonders, is a sight so grand and novel as to well repay the exposure and fatigue of a visit thither through bleak January's snows and ice.

The Mountain House is then closed, but Mr. Scutt inhabits his Laurel Inn all the year, we believe. This is a hint to the enthusiast in the search for the strange and beautiful in Nature. Most tourists will care to see the Catskills only in July or August.

Charges at the Mountain House are, as in most of the fashionable summer resorts in the United States, $2.50 per day. At the Laurel Inn, by the High Falls, about half that price, we believe.

Stages will take you back to Catskill village, as they have brought you thence, in season for steamboats and railways, for elsewhere.

To visit the Catskills comfortably, three days will suffice for the journey thence by rail from New York, for the stay and the return to the city. Not less than four, however, ought to be thus invested, if one would make sure of a satisfactory dividend; and if a week is at command, so much the happier he who commands it.

NEW YORK TO ALBANY.
Via Harlem Railroad.

This Route extends from the heart of the city of New York to the State capital, skirting in its course the eastern portions of all those counties lying upon the Hudson and traversed by the river railway. The distance between the termini is 154 miles, a few miles longer than that of the Hudson River. Time, about the same. The stations and towns upon the Harlem Road are, for the most part, inconsiderable places, many of them having grown up with the road. The country passed through is varied and picturesque in surface, and much of it is rich agricultural land. It does not compare with the river route in scenic attractions.

STATIONS.

New York—corner of White and Centre street—Yorkville, 6 miles; Harlem, 7; Mott Haven, 8; Melrose, 9; Morrisania, 10; Tremont, 11; Fordham, 12; William's Bridge, 14; (Junction of the N. Y. and N. Haven Road) Hunt's Bridge, 16; Bronxville, 18; Tuckahoe, 19; Scarsdale, 22; Hart's Corners, 24; WHITE PLAINS, 26; Kensico, 29; Unionville, 31; Pleasantville,

34; Chapequa, 36; Mount Kisco, 40; Bedford, 42; Whitlockville, 45; Golden Bridge, 47; Purdy's, 49; Croton Falls, 51; Brewster's, 55; Dykeman's, 58; Towners, 61; Patterson's, 63; Pawlings, 67; South Dover, 73; Dover Furnace, 76; Dover Plains, 80; Wassaic, 84; Amenia, 88; Sharon Station, 91; Millerton, 96; Mount Riga, 99; Boston Corners, 103; Copake, 108; Hillsdale, 112; Bains, 115; Martindale, 118; Philmont, 122; Ghent, 128; CHATHAM FOUR CORNERS, 130; (Junction with railway route from Albany and from Hudson, for Boston) EAST ALBANY, 153 miles.

All the Stations from New York to White Plains (26 miles) are suburban, being escape valves of the over-grown population of the city, where the business of the principal part of their population lies, and to which they go daily by the railway. Many of the villages are picturesque, pleasant and prosperous.

On leaving the city streets, the road passes under a considerable extent of tunnelling and continued bridging across thoroughfares overhead, making merry diversion for the passengers. At the extremity of the Island and city of New York at Harlem, the road crosses the Harlem River into Westchester County.

White Plains (Westchester County), is interesting as the scene of important events in the Revolution. An eventful battle was fought here, Oct. 28, 1776. A residence of Washington (in which are some attractive relics) is yet standing in the vicinage.

Croton Falls, upon the river which supplys the great Croton Aqueduct to the city of New York.

Lake Mahopac. Passengers for Lake Mahopac take stage thence (distance two hours) at the Croton Falls Station. See "Lake Mahopac."

Dover Plains, 20 miles east of Poughkeepsie, is surrounded by much pleasing landscape.

For Albany, and routes thence by Hudson River and the river railway, see Index.

SARATOGA SPRINGS.

From Boston by the Western Railway, 200 miles to Albany; or, from New York by the Hudson River line or steamboats, 144 miles to Albany. Thence 32 miles by the Rensselaer and Saratoga Road, through the city of Troy and Ballston Springs. Journeying from New York by water, the traveller may pass pleasantly up the river, during the night, breakfast in Albany, and reach the Springs in good season next morning. The little ride from Albany to the Springs is a most agreeable one, as the route crosses and follows the Hudson and the Mohawk Rivers, as it passes Waterford at the meeting of these waters, four miles above Troy, and near the Cohoes Falls, a much-admired and frequented resort upon the Mohawk, as it thence continues upon the west bank of the Hudson, eight miles further to Mechanicsville. It afterwards crosses the canal, passes Round Lake, and enters Ballston Springs.

Ballston Spa is upon the Kayaderosseros Creek, a small stream which flows through the village, 25 miles from Troy. Its mineral waters, which were discovered in 1769, are celebrated for their medicinal qualities, although not so popular as they were formerly, those of Saratoga being now generally preferred. The *Sans Souci Hotel* is a pleasant house near the centre of the village.

Five miles distant is Long Lake, a resort of the angler. Saratoga Lake is six miles from Ballston.

Saratoga has been for many years, and still is, and probably always will be, the most famous place of summer resort in the United States, frequented by Americans from all sections, and by foreign tourists from all climates. During the height of the fashionable season no less than two or three thousand arrivals occur within a week. There is nothing remarkable about the topography or the scenery of Saratoga; on the contrary, the spot would be uninteresting enough but for the virtues of its waters and the pleasures of its brilliant society. The village streets, however, are gratefully shaded by fine trees, and a little "let up" in the gay whirl may be got on the walks and lawns of the pretty rural cemetery close by.

The most desirable hotels at Saratoga are the United States, the Union Hall and Congress Hall. Besides this famous trio of houses, there are many of less fashion and price, besides numerous private boarding-houses, where one may live quietly at a moderate cost. The three hotels which we have named have each accommodations for six or seven hundred guests, for all of which the demand is more than ample. Fine bands of music discourse on the broad, shady piazzas, and in the ball-rooms, at the dinner and evening hours. The Clarendon is a new spacious house, in process of erection as we now write.

The health-giving Springs of which the fame of Saratoga has been born, however much Fashion may have since nursed it, are all in or very near the village. There are many different waters in present use, but the most sought after of all are those of the Congress Spring, of which Dr. Chilton gives us an analysis thus:—One gallon, of 231 cubic inches—chloride of sodium, 363.829 grains; carbonate of soda, 7.200; carbonate of lime, 86,143; carbonate of magnesia, 78.621; carbonate of iron, .841; sulphate of soda, .651; iodine of sodium and bromide of potassium, 5.920; silica, .472; alumina, .321; total, 543.998 grains. Carbonic acid, 284.65; atmospheric air, 5.41: making 290.06 inches of gaseous contents.

This Spring was discovered in 1792, though it was long before known to and esteemed by the Indians.

After the Congress waters, which are bottled and sent all over the world, as everybody knows, the Springs most in favor and use at Saratoga are the Empire, the Columbian, the High Rock, the Iodine, the Pavilion and Putnam's. The Empire Spring, the most northerly one in the village, has grown greatly in repute of late years. So far its landscape surroundings have received but

Congress, Spring, Saratoga.

poor attention. The High Rock Spring, not far from the Empire, is much esteemed both for its medical virtues and for the curious character of the rock from which it issues, and after which it is named. This singular rock has been formed by the accumulated deposits of the mineral substances (magnesia, lime and iron) held in solution by the carbonic acid gas of the Springs. The circumference of the rock, at the surface of the ground, is 24 feet 4 inches, its height $3\frac{1}{2}$ feet, with an aperture of nearly one foot diameter.

The Alpha and the Omega of the daily Saratoga programme, is to drink and to dance—the one in the earliest possible morning, and the other at the latest conceivable night. Among the outside diversions is a jaunt to Saratoga Lake, a pleasant water six miles away. Here they have nice boating fun, and sometimes "make believe" to fish. This lake is nine miles in length and very near three in width. The marshes around it prevent access, except here and there. *Snake Hill* steps into the water, and lifts up its head 40 feet or so, upon the eastern side of the lake.

"There is," so the story has been told, "an Indian superstition attached to this lake, which probably had its source in its remarkable loneliness and tranquillity. The Mohawks believed that its stillness was sacred to the Great Spirit, and that if a human voice uttered a sound upon its waters, the canoe of the offender would instantly sink. An Englishwoman once, in the early days of the settlers, had occasion to cross this lake with a party of Indians, who, before embarking, warned her, most impressively, of the spell. It was a silent, breathless day, and the canoe shot over the smooth surface of the lake like a shadow. About a mile from the shore, near the centre of the lake, the woman, willing to convince the savages of the weakness of their superstition, uttered a loud cry. The countenances of the Indians fell instantly to the deepest gloom. After a moment's pause, however, they redoubled their exertions, and, in frowning silence, drove the light bark like an arrow over the waters. They reached the shore in safety, and drew up the canoe, and the woman rallied the chief upon his credulity. 'The Great Spirit is merciful,' answered the scornful Mohawk, ' he knows that *a white woman cannot hold her tongue.*' "

A visit to Lake George, 28 miles distant by rail and plank road, is a delightful episode and variation in Saratoga life. See Lake George.

LAKE GEORGE.

The route from New York, Boston, and the West, to Lake George, is through Saratoga, and thus far is the same as to that point; thence to Moreau station, 15 miles, by the Troy and Whitehall line, and from there to Caldwell, at the south end or head of the lake by plank road.

Glen's Falls, in the Upper Hudson, is on the way, nine miles from the lake. The wild and rugged landscape is in striking contrast with the general air of the country below—there, quiet pastoral lands; here, rugged rock and rushing cataract. This is a spot trebly interesting, from its natural, its poetical, and its historical character. The passage of the river is through a rude ra-vine, in a mad descent of 75 feet over a rocky precipice 900 feet in length. Within the roar of these cataracts were laid some of the scenes in Cooper's story of the "Last of the Mohicans." They are gently associated with our romantic memories of Uncas and Hawk's Eye, David, Duncan Haywood and his sweet wards, Alice and Cora Monroe.

When within four miles of the lake, we pass a dark glen, in which lie hidden the storied waters of *Bloody Pond,* and close by is the historic old boulder remembered as *Williams' Rock.* Near this last-named spot, Colonel Williams was killed in an engagement with the French and Indians, Sept. 8, 1755. The slain in this unfortunate battle were cast into the waters near by, since called Bloody Pond. It is now quiet enough, under its surface of slime and dank lilies.

The first broad view of the beautiful lake, seen suddenly as our way brings us to the brink of the high lands, above which we have thus far travelled, is of surpassing beauty, scarcely exceeded

Lake George, N. Y.

by the thousand-and-one marvels of delight which we afterwards enjoy in all the long traverse of the famous waters.

Our road now descends to the shore, the gleaming floods and the blue cliffs of Horicon still, ever and anon, filling our charmed eye. We halt at the Lake House at Caldwell village, or at the Fort William Henry Hotel, a new and elegant establishment near by, at the ruins of the Old Fort, on the right.

About a mile south-east from the site of Fort William Henry are the ruins of Fort George. These localities are seen from the piazza of the Lake House, which commands also a fine view of the French Mountain and Rattle-Snake Hill, and of the islands and hills down the lake.

The passage of Lake George, 36 miles, to the landing near the village of Ticonderoga, and four miles from the venerable ruins of Fort Ticonderoga, on Lake Champlain, is made by steamboat,* the trip down to the Fort and back occupying the day very delightfully.

Leaving Caldwell after breakfast, we proceed on our voyage down the lake. The first spot of especial interest which we pass is Diamond Island, in front of Dunham Bay. Here, in 1777, was a military depot of Burgoyne's army, and a skirmish between the garrison and a detachment of American troops.

North of Diamond Isle, lies Long Island, in front of Long Point, which extends into the lake from the east. Harris Bay lies between the north side of this Point and the mountains. In this bay Montcalm moored his boats and landed, in 1757.

Dome Island is passed, in the centre of the lake, some 12 miles north of Caldwell. Putnam's men took shelter here while he went to apprise General Webb of the movements of the enemy, at the mouth of the North-West bay. This bay lies in one of the most beauti-

ful parts of Lake George, just beyond Bolton Landing, where there is an inviting place of sojourn called the "Mohican House." The bay extends up on the west of the Tongue Mountain some five miles. On the east side of the bay the Tongue Mountain comes in literally like a tongue of the lake, into the centre of which it seems to protrude, with the bay on one side and the main passage of the waters on the other. On the right or east shore in the neighborhood, and just as we reach the Tongue and enter the "Narrows," is the bold semicircular palisades called Shelving Rock. Passing this picturesque feature of the landscape, and, afterwards, of the point of the Tongue Mountain, we enter the NARROWS at the base of the boldest and loftiest shores of Horicon. The chief peak of the hills here, is that of Black Mountain, with an altitude of 2,200 feet. The islands are numerous, though many of them are merely peeps out of the water. The best fishing-grounds of Lake George are in that part of the waters which we have already passed, in the vicinity of Bolton Landing, Shelving Rock, and thence to Caldwell, though fine trout and bass are freely caught from one end of the lake to the other.

Sabbath Day Point. Emerging from the Narrows, on the north, we approach a long projecting strip of fertile land, called Sabbath Day Point— so named, by General Abercrombie, from his having embarked his army on the spot on Sunday morning, after a halt for the preceding night. The spot is remembered, also, as the scene of a fight, in 1756, between the colonists and a party of French and Indians. The former, sorely pressed, and unable to escape across the lake, made a bold defence and defeated the enemy, killing very many of their men. Yet, again, in 1776, Sabbath Day Point was the scene of a battle between some American militia and a party of Indians and Tories, when the latter were repulsed, and some 40 of their number were killed and wounded. This part of Horicon is even more charming in its

---

* The "John Jay," which has plied the waters of Lake George for some years past, was destroyed by fire in the summer of 1856, opposite Garfield's, near Sabbath Day Point.

CANADA EAST

La Prairie 9 miles from Montreal
St. Johns, Amherst, Montreal 37, from New York and 306 from Boston

Burtonville
Montreal R.
La Cole
Rouses
Corbeau
Chazy Land?
I. la Mott
Ram's Head
Cumberland Head
PLATTSBURG 103 m. from Whitehall & 161 fr. Albany
P. Jackson
Peru Land?
Port Kent
Keesville
Pt. Douglass
Schuyler
Peru Bay
Wills borough
Essex
Split Rock
West port 55
Port Henry
Crown
Putnam C.
Mt. Hope
Ticonderoga 124
Alexandria
Independence

St. Johns
Amherst
Granby
N.E. Branch
Sheffield
L. Broom
L. Tome fobis
orgeville
Stanste

Pike R.
Dunham
Memphremagog Lake
Phillipsburg
Highgate
N. Troy
Missisque R.
Swanton
Montgomery
W. Chan
IRASBU
Bakersfield
ORLE
St. Albans
Georgia
Waterville
Eden
Barton
Be water
La Moille R.
Cambridge
Graftsbury
Sutton
West Milton
Johnston
N.
Colchester
Underhill
Greensboro
Jericho
Elmore
CALEDONIA
Stowe
DANVILLE
Williston
Winooski
Bolton
Waterbury
Cabot
Charlotte
Onion River
Groton
Ferrisburg
MONTPELIER 58 m. from Lebanon 27 fr. Concord and 203 from Boston
Barre
Wells R.
Waitsfield
Orange
Vergennes
Northfield
N. Ha.
Addison
New Haven
Roxbury
Newbury
Port Henry
Chimney Pt.
MIDDLEBURY
ORANGE
Haverh
Bridport
CHELSEA
Bradford
Shoreham
Trout Pd.
Randolph
Rochester
Bethel
Orford
Brandon
Thetford
Lyme
Alexandria
Barnard
Norwich
Dartmouth College
Pittsford
Royalton
WINDSOR
Hartford
Hanover
Bomba zine L.
WOOD STOCK
70 m. from Concord
NORTHERN R.R.
RUTLAND from Boston via Fitchburg 167 miles
W. Lebanon

WORK
FRANKLIN
GRAND ISLE
S. Hero
N. Hero
CHITTENDEN
LAMOILLE
WASHINGTON
VERMONT
ADDISON
RUTLAND
ORANGE
WINDSOR
CONNECTICUT RIVER
NEW HAMPSHIRE

LAKE GEORGE
Mt. Defiance
Shelburne
Burlington
Fair Haven 232
Boston 233

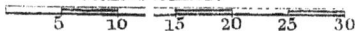

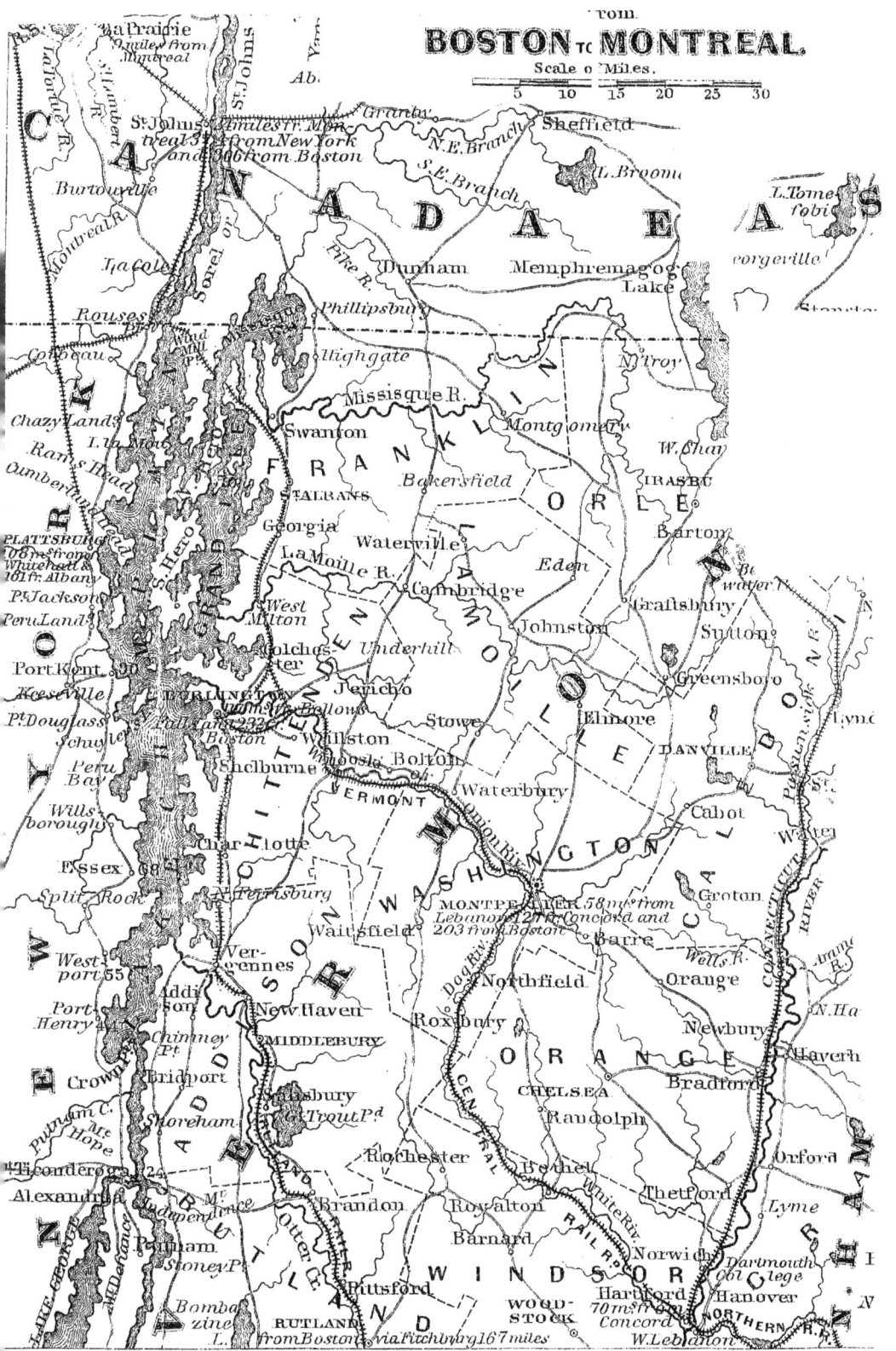

pictures, both up and down the lake, than it is in its numerous historical reminiscences. On a calm sunny day the romantic passage of the Narrows, as seen to the southward, is wonderfully fine; while, in the opposite direction, the broad bay, entered as the boat passes Sabbath Day Point, and the summer landing and hotel at "Garfield's," are soon to be abruptly closed on the north by the huge precipices of Anthony's Nose on the right, and Rogers' Slide on the left. This pass is not unlike that of the Highlands of the Hudson as approached from the south.

Rogers' Slide is a rugged promontory, about 400 feet high, with a steep face of bare rock, down which the Indians, to their great bewilderment, supposed the bold ranger, Major Rogers, to have passed, when they pursued him to the brink of the precipice. Two miles beyond is *Prisoner's Island,* where, during the French war, those taken captive by the English were confined; and directly west is *Lord Howe's Point,* where the English army, under Lord Howe, consisting of 16,000 men, landed previous to the attack on Ticonderoga. We now approach the termination of our excursion on this beautiful lake, and in a mile reach the steamboat landing near the village of *Ticonderoga,* whence stages run a distance of three miles, over a rough and romantic road, to *Fort Ticonderoga*—following the wild course of the passage, by which Horicon reaches the waters of Lake Champlain—a passage full of bold rapids and striking cascades.

After exploring the picturesque ruins of the ancient fort, and dining satisfactorily at the excellent hotel, which stands upon the marge of a beautiful lawn, sloping to the Champlain shore, our stage will take us back to the landing we have left on Lake George, and our steamboat thence to Caldwell again, in time for tea and a moonlight row among the countless green isles; or we may take the Champlain boat to Whitehall; or from Whitehall *en route* for Canada.

7*

Fort Ticonderoga, of which the ruins only are visible, was erected by the French in 1756, and called by them "Carrillon." It was originally a place of much strength; its natural advantages were very great, being surrounded on three sides by water, and having half its fourth covered by a swamp, and the only point by which it could be approached, by a breastwork. It was afterwards, however, easily reduced, by an expedient adopted by General Burgoyne—that of placing a piece of artillery on the pinnacle of *Mount Defiance,* on the south side of the Lake George outlet, and 750 feet above the lake, and entirely commanding the fort, from which shot was thrown into the midst of the enemy's works. Fort Ticonderoga was one of the first strongholds taken from the English in 1775, at the commencement of the Revolutionary war. Colonel Ethan Allen, of Vermont, at the head of the Green Mountain Boys, surprised the unsuspecting garrison, penetrated to the very bedside of the commandant, and waking him, demanded the surrender of the fort. "In whose name, and to whom?" exclaimed the surprised officer.—"In the name of the great Jehovah, and the Continental Congress!" thundered the intrepid Allen, and the fort was immediately surrendered.

NEW YORK TO MONTREAL AND NIAGARA, via LAKE CHAMPLAIN.

One of the most delightful of American summer tours is from New York, via the Hudson River, Lake Champlain, and the St. Lawrence River, to the Falls of Niagara, returning by the lower routes—the Central or the Erie Railways. From Boston and Portland, lines of railroad connect conveniently with the St. Lawrence routes. A thousand places and objects of interest fall within the direct line of this journey; besides which, it has many alluring asides, which may be readily reached.

From Portland, Maine, take the Grand Trunk Route, to Montreal (or to Quebec)—from Boston, take the Boston, Concord, and Montreal Routes.

From New York take the Hudson River Route, which we have already travelled, to Albany and Troy; thence, by rail, via Saratoga Springs, to Whitehall, at the southern extremity of Lake Champlain. We resume the programme at Moreau Station, on this line, to which point we have already followed it in our visit to Lake George. At Ticonderoga, above, on Lake Champlain, we shall meet those who prefer, as many do, to pursue the journey to that point, by the way of Lake George, instead of via Whitehall and the lower end of Champlain.

To Whitehall the country is exceedingly attractive, much of the way, in its quiet, sunny, valley beauty, watered by pleasant streams, and environed, in the distance, by picturesque hills. The Champlain Canal is a continual object of interest by the way; and there are, also, as in all the long journey before us, everywhere spots of deep historic charm, if we could tarry to read their stories—of the memorable incidents which they witnessed, both in the French and Indian, and afterwards in the Revolutionary war. In the valley regions of the Hudson, which lie between Albany and Lake Champlain, are many scenes famous for the struggles between the Colonists and Great Britain —the battle-grounds of Bemis Heights and Stillwater (villages of the Upper Hudson), and of Saratoga, which ended in the defeat of Burgoyne and his army. Then there is the tale of the melancholy fate of Jane M'Crea, so cruelly murdered by the Indians at Fort Edward; and many histories, which it is pleasant to recall ever so vaguely, as we pass along.

Whitehall was a point of much consideration during the French and Indian war, and through the Revolution. In former times it was called Skenesborough. It is at the south end or the head of Lake Champlain, within a rude rocky ravine, at the foot of Skene's Mountain. Its position, as a meeting-place of great highways of travel, has made it quite a bustling and prosperous village. There is nothing

Lake Champlain, N. Y.

NEW YORK. 155

Lake Champlain—The Walled Banks of the Ausable—Battle at Plattsburg.

in the vicinage, however, to delay the traveller. From Whitehall our journey lies down Lake Champlain, 156 miles, to St. Johns, though we might, instead, go by railway through Vermont, via Castleton, Rutland, Burlington, &c., to Rouse's Point, and thence, still by railway, to Montreal.

The narrowness of the lower part of Lake Champlain gives it much more of a river than lake air. For 20 miles the average breadth does not exceed half a mile; and, at one point, it is not more than 40 rods across. However it grows wide enough as we pass Ticonderoga,* where passengers by the Lake George *détour* are picked up, and in the vicinity of Burlington there are too many broad miles between the shores for picturesque uses. Whether it is broad though or narrow, the voyage, in large and admirable boats, over its mountain-environed waters, is always a pleasure to be greatly enjoyed and happily remembered. On the east rise the bare peaks of the Green Hills of Vermont, the bold Camel's Hump leading all the long line; and on the west are the still more varied summits and ridges of the Adirondack Mountains in New York.

Mount Independence lies in Vermont, opposite Ticonderoga, about a mile distant. The remains of military works are still visible here.—*Mount Hope*, an elevation about a mile north from Ticonderoga, was occupied by General Burgoyne previous to the recapture of Ticonderoga, which took place in 1777, nearly two years after its surrender to the gallant Allen. St. Clair, the American commander, being forced to evacuate, it again fell into the possession of the British, and was held during the war.

Not far above, and upon the opposite shore, is the village of Crown Point; and, just beyond, the picturesque and well-preserved ruins of the fortifications of the same name. Opposite is Chimney Point; and, just above, on the left, at the mouth of Bulwaggy Bay, is Port Henry.

* See Lake George for " Ruins of Fort Ticonderoga.

Burlington, the largest town on the lake, is upon the eastern or Vermont shore, about midway between Whitehall and St. Johns. Rising gradually to an elevation of several hundred feet, it is imposingly seen from the water. It is the seat of the University of Vermont, and is a place of much commercial importance, connected by railways with all parts of the country. Across the lake is

Port Kent, from which vicinity, whether on land or on water, the landscape in every direction is exceeding striking and beautiful.

The Walled Banks of the Ausable. The remarkable Walled Banks of the Ausable are a mile or two west of Port Kent, on the way to the manufacturing village of Keeseville.

It is at the Ausable House, an excellent summer hotel in the picturesque village of Keeseville, that the traveller will establish himself, if he would visit this wonderful ravine, with its grand walls and its rushing waters. The Falls of the Ausable, though they are but little known as yet, will one day be esteemed among the chief natural wonders of the country.

Plattsburg. Above and opposite Burlington is the pleasant village of Plattsburg, where the Saranac river comes in from its lake-dotted home, at the edge of the great wilderness of northern New York, 30 miles westward.

Battle of Lake Champlain. Plattsburg was the scene of the victory of M'Donough and Macomb over the British naval and land forces, under Commodore Downie and Sir George Provost. Here the American commodore awaited at anchor the arrival of the British fleet, which passed Cumberland Head about eight in the morning of the 11th September, 1814. The first gun from the fleet was the signal for commencing the attack on land. Sir George Provost, with about 14,000 men, furiously assaulted the defences of the town, whilst the battle raged between the fleets, in full view of the armies. General Macomb, with about

Trenton Falls, N. Y.

3,000 men, mostly undisciplined, foiled the repeated assaults of the enemy; until the capture of the British fleet, after an action of two hours, obliged him to retire, with the loss of 2,500 men and a large portion of his baggage and ammunition. Here we might land and take the Plattsburg and Montreal Railway, 62 miles direct to Montreal.

Rouse's Point, on the west side of the lake, is the last landing-place before we enter Canada. Railways from the Eastern States, through Vermont, come in here, and are prolonged by the Champlain and St. Lawrence road to Montreal. If the traveller towards Canada continues his journey, neither via Plattsburg nor Rouse's Point, he may go on by steamboat to the head of navigation on these waters to St. Johns, and thence by Lachine to Montreal.

See Canada for the tour of the St. Lawrence and Lake Ontario from Montreal to Niagara.

NEW YORK TO TRENTON FALLS.

Via Hudson River to Albany, thence by the New York Central Railroad as far as Utica, and thence, over a plank road, or by Rail Road, 15 miles.

Trenton Falls, says Mr. Willis, "is the most *enjoyably beautiful* spot among the resorts of romantic scenery in our country. The remembrance of its loveliness becomes a bright point, to which dream and reverie oftenest return. It seems to be curiously adapted to enjoy, being somehow, not only the *kind* but the *size* of a place which the (after all) *measurable* arms of a mortal heart can hold in its embrace. Niagara is too much, as a roasted ox is a thing to go and look at, though one retires to dine on something smaller."

Trenton Falls is the place, above all others, where it is a luxury to *stay*— which one oftenest revisits, which one most commends to strangers to be sure to see.

"In the long corridor of travel between New York and Niagara, "Trenton, Mr. Willis says again, "is a sort of alcove aside—a side-scene out of earshot of the crowd—a recess in a window, whither you draw a friend by the button for the sake of chit-chat at ease."

Trenton Falls is rather a misnomer, for the wonder of nature which bears the name is a tremendous torrent, whose bed, for several miles, is sunk fathoms deep into the earth—a roaring

Trenton Falls.

and dashing stream, so far below the surface of the forest, in which it is lost, that you would think, as you come suddenly upon the edge of its long precipice, that it was a river in some inner world (coiled within ours, as we in the outer circle of the firmament), and laid open by some Titanic throe that had cracked clear asunder the crust of this " shallow earth." The idea is rather assisted if you happen to see below you, on its abysmal shore, a party of adventurous travellers; for at that vast depth, and in contrast with the gigantic trees and rocks, the same number of well-shaped pismires, dressed in the last fashion, and philandering upon your parlor floor, would be about of their apparent size and distinctness.

Trenton Falls are upon the West Canada Creek, a branch of the Mohawk. The descent of the stream, 312 feet in a distance of two miles, is by a series of half a dozen cataracts, of wonderful variety and beauty. Every facility of path and stairway, and guide, for the tour of the Trenton ravine has been provided by Mr. Moore, who has for many years resided on the spot, and been always its Prospero, and its favorite host.

A walk of a few rods through the woods brings the visitor to the brink of the precipice, descended by secure stairways for some hundred feet.

The landing is a broad pavement, level with the water's edge, often, in times of freshet, the bed of foaming floods. Here is commanded a fine view of the outlet of the chasm, 45 rods below, and also of the first cascade, 37 rods up the stream.

The parapet of the First Fall, visible from the foot of the stairs, is, in dry times, a naked perpendicular rock, 33 feet high, apparently extending quite across the chasm, the water retiring to the left and being hid from the eye by intervening prominences. But in freshets, or after rain, it foams over, from the one side of the gorge to the other, in a broad amber sheet. A pathway to this fall has been blasted at a considerable cost, under an overhanging rock and around an extensive projec- tion, directly beneath which rages and roars a most violent rapid. The passage, though at first of dangerous aspect, is made secure by chains well riveted in the rocky wall. In the midst of this projection, five tons were thrown over by a fortunate blast, affording a perfect- ly level and broad space, where 15 or 20 persons may find ample footing, and command a noble view of the entire scene. A little to the left, the rapid commences its wild career. Directly underneath, it rages and foams with great fury, forcing a tortuous passage into the expanded stream on the right. In front is a projection from the other side, curved to a concavity of a semi-circle by the impetuous waters. The top of this projection has been swept away, and is entirely flat, exhibiting from its surface downwards, the sepa-rate strata as regular and distinct, and as horizontal as mason-work in the lock of the grand canal. Here, in the old time, was a lofty fall, now reduced to the rapid we have described.

Beyond, massive rocks, thrown over in flood times, lie piled up in the mid-dle of the river. Passing to the left, yet a few rods above, we come into the presence of Sherman's Fall, so named in memory of the Rev. Mr. Sherman, whose account of the spot we are now closely following. He was one of the earliest pioneers of the Trenton beau-ties, and it was by him that the first house, called the " Rural Resort," for the accommodation of visitors, was built. It has formed an immense ex-cavation, having thrown out thousands of tons from the parapet rock, visible at the stairs, and is annually forcing off slabs at the west corner, against which it incessantly forces a section of its powerful sheet. A naked mass of rock, extending up 150 feet, juts frown-ingly forward, which is ascended by natural steps to a point from which the visitor looks securely down upon the rushing waters.

Leaving this rocky shelf, and passing a wild rapid, we come suddenly in sight of the High Falls, 40 rods beyond.—

Trenton Falls.

Trenton Falls, N. Y.

This cascade has a perpendicular descent of 100 feet, while the cliffs on either side, rise some 80 feet yet higher. The whole body of water makes its way at this point—divided by intervening ledges into separate cataracts, which fall first about 40 feet, then reuniting on a flat below, and veering suddenly around an inclination of rocky steps, they plunge into the dark caldron beneath.

Passing up at the side, we mount a grand level, where in dry times the stream retires to the right and opens a wide pavement for a large party to walk abreast. Here a flight of stairs leads to a refreshment house, called the Rural Retreat, 20 feet above the summit of the High Falls.

The opening of the chasm now becomes considerably enlarged, and a new variety of scene occurs. Mill Dam Fall, 14 feet high, lies some distance beyond, reaching across the whole breadth of the chasm.

Ascending this Fall the visitor comes to a still larger platform of level rock, 15 rods wide at low water, and 90 in length, lined on each side by cedars. At the extremity of this locality, which is known as the Alhambra, a bare rock 50 feet in height reaches gradually forward from the mid-distance; and, from its shelving top, there descends a perpetual rill, which forms a natural shower-bath. A wild cataract fills the picture on the left.

Here the wide opening suddenly contracts and a narrow aperture only remains, with vistas of winding mountain, cliff and crag. Near by is a dark basin, where the waters rest from the turmoil of the wild cascade above. In this vicinage is an amphitheatre of seemingly impossible access, replete with even new surprises aad delights. Yet beyond, is the Rocky Heart, the point at which the traverse of the ravine usually ends, though despite the difficulties and dangers of the way, even ladies frequently penetrate beyond as far as the falls at Boon's Bridge, the terminus of the gorge.

The scene at Trenton varies much, according as drought or freshet dries or fills the stream, and passages are easy enough at one time, which are utterly impracticable at others. It is difficult to say when the glen is the most beautiful, whether with much or with little water.

Trout once inhabited these waters, but they are gone now. Game, too, is scarce in the vicinage, though partridges, wild ducks, snipes, black and gray squirrels, woodcock and the rabbit may yet be taken. Trenton is a spot for a long sojourn, though it may be run over pleasantly in a day.

NEW YORK TO BUFFALO.

To Albany by the Hudson River, 146 miles, and thence by the New York Central—a chain of railways 298 miles.

This great route traverses, from east to west, the entire length of the Empire State. It has two termini at the eastern end, one at Albany, and the other at Troy, which meet, after 17 miles, at Schenectady. It then continues, in one line, to Syracuse, 148 miles from Albany; when it is again a double route for the remainder of the way; the lower line being looped up to the other about midway, between Syracuse and Buffalo, at Rochester. The upper route is the more direct and the one which we shall now follow. The great Erie Canal traverses the State of New York from Albany to Buffalo, nearly on the same line with the Central Railroad.

Trains leave Albany and Troy for Buffalo and all points west to the Mississippi and beyond, on the arrival there of the cars from the south, east and north—New York, Boston and Canada.

At **Schenectady** the railways from Albany and Troy meet, and the Saratoga route diverges. Schenectady is upon the bank of the Mohawk. It is one of the oldest towns in the State, and is distinguished as the seat of *Union College*. The council-grounds of the Mohawks were once on this spot. In the winter of 1690, a party of two hundred Frenchmen and Canadians, and fifty Indians, fell at midnight upon Schenectady, killed and made captive its people, and burned the village to ashes. Sixty-nine persons were then massacred and twenty-seven were made prisoners. The church and sixty-three houses were destroyed. It was afterwards taken in the French war of 1748, when about seventy people were put to death.

Leaving Schenectady, the road crosses the Mohawk River and the Erie Canal, upon a bridge nearly one thousand feet in length.

At **Palatine Bridge,** 55 miles from Albany, passengers for Sharon Springs leave the road and proceed by stage. See Sharon Springs.

At **Fort Plain,** 68 miles from Albany, passengers for Otsego Lake, Cooperstown and Cherry Valley, proceed by stage.

Little Falls is remarkable for a bold passage of the Mohawk River and the Erie Canal through a wild and most picturesque defile. The scenery, with the river rapids and cascades, the locks and windings of the canal, the bridges, and the glimpses, far away, of the valley of the Mohawk, is especially beautiful.

At **Utica,** 95 miles from Albany, a railway and canal come in from Binghamton, on the line of the Erie Road. Here passengers leave for Trenton Falls (see Trenton Falls), 15 miles distant. Utica is a large and thriving place, with many fine public and private buildings. It is built upon the site of old Fort Schuyler, and has now a population of over 22,000. *Hotels.*—Bagg's, connected with the Railway Depôt.

At **Syracuse,** 148 miles from Albany, the Central Road connects by rail with Binghamton on the Erie Route, and with Oswego, northward. The most extensive salt manufactories in the United States are found here. It is famous, too, as the meeting-place of State political and other conventions. Syracuse is a large and elegant city, with a population of over 26,000.— *Hotels.*—The Globe, the Syracuse and the Onondaga.

Auburn. This important city is near Owasco Lake, a beautiful water, 12 miles long. It is the seat of the Auburn State Prison.

Skaneateles is five miles distant, by a branch railway, at the foot of Skaneateles Lake, a charming water, 16 miles long, with picturesque shores and good supplies of trout and other fish.

Cayuga is a pleasant village upon the eastern shore of Cayuga Lake.— *Ithaca* is 38 miles off at the other extremity of the Lake. These fine waters are traversed daily by steamboat, connecting Cayuga with Ithaca, and by railway with Oswego, on the New York and Erie route.

Geneva is upon Seneca Lake, one of the largest and most beautiful of the many lakes of western New York. It is 40 miles long and from two to four wide. Steamboats connect its towns and villages with the great routes of travel. The Hobart Free College, under the direction of the Episcopalians, is here; also the Medical Institute of Geneva College and the Geneva Union School.

Canandaigua is a beautiful village, at the north end of Canandaigua Lake, The railroad from Elmira, on the New York and Erie route, to Niagara Falls, passes through Canandaigua. The lake is about 15 miles in length, and is well stocked with fish.

Rochester is the largest and most important city upon our present route, between Albany and Buffalo, its population being nearly 45,000. It is the seat of the Rochester University, founded by the Baptists in 1850. There is also here a Baptist Theological Seminary, founded in 1850. Among its picturesque attractions, are the Falls of the Genesee, upon both sides of which river the city is built. The Mount Hope Cemetery in the vicinity, is also a spot of much natural beauty.

Rochester is connected by railway with the New York and Erie route at Corning, and with Niagara Falls direct, by the Rochester, Lockport, and Niagara Falls Division of the New York Central Road, and by steamboats, with all ports on Lake Ontario.

Hotels.—The Eagle and the Congress Hotels, are among the many excellent houses here.

The **Genesee Falls** are seen to the best advantage from the east side of the stream. The railroad cars pass about 100 rods south of the most southerly fall on the Genesee River, so that passengers in crossing lose the view. These falls have three perpendicular pitches, and two rapids; the first great cataract is 80 rods below the aqueduct, the stream plunging perpendicularly 96 feet. The ledge here recedes up the river from the centre to the sides, breaking the water into three distinct sheets.

From *Table Rock*, in the centre of these falls, Sam Patch made his last and fatal leap. The river below the first cataract is broad and deep, with occasional rapids to the second fall, where it again descends perpendicularly 20 feet. Thence the river pursues its course, which is noisy, swift, and rapid, to the third and last fall, over which it pours its flood down a perpendicular descent of 105 feet. Below this fall are numerous rapids which continue to Carthage, the end of navigation on the Genesee River from Lake Ontario.

At Rochester the two routes of the Central Road unite, and again diverge to reunite at Buffalo. By the upper route the traveller will pass through Lockport direct to Niagara, leaving Buffalo to the south-west. The lower route, direct to Buffalo, is intersected at Batavia, by the Buffalo and Corning Road, from Corning on the Erie Railway, via Rochester to Niagara.

Buffalo. We have now reached the shores of Lake Erie, and are at the end of our route, whence we may proceed at our pleasure, by steamboat or railway, to any place northward or southward, in the Far West; for Buffalo is the point where routes of travel most do meet.

This important commercial and manufacturing city has grown so great and so fast, that although it was laid out as late as 1801, and in 1813 had only 200 houses, its population now numbers nearly 80,000. It is an earnest of the wonderful progress which we shall see by-and-by, when we continue our

Routes to the Falls of Niagara.

travels hence, towards the further West. Stop at the American.

NIAGARA FALLS.

ROUTES.—From *New York*, via Hudson River and Hudson River Railroad, to Albany, 146 miles; from Albany to Buffalo, via N. Y. Central R. R., 298 miles; from Buffalo, by Buffalo, Niagara Falls, and Lewistown R. R. (to Niagara), 22 miles. Total, 466 miles.

From *New York*, via New York and Erie R. R., to Buffalo, 422 miles; Buffalo (as above), by Buffalo, Niagara Falls, and Lewistown R. R. (to Niagara), 22 miles. Total, 444.

From *New York*, by New York and Erie R. R. to Elmira, 273 miles; from Elmira to Niagara, by Elmira, Canandaigua, and Niagara Falls R. R., 166 miles. Total, 439 miles.

Passengers can leave the main N. Y. Central Railway (from Albany to Buffalo) at Rochester, and take the Rochester, Lockport, and Niagara division, 76 miles, thence to Niagara.

From *New York* to *Albany*, by Hudson River, 146 miles; thence to Troy, 6 miles. Railway from Troy to Whitehall, sixty-five miles; from Whitehall by steamer on Lake Champlain, to St. Johns, 150 miles; .St. Johns to La Prairie Railroad, 15 miles; La Prairie, steamboat on the St. Lawrence to Montreal, 9 miles; from Montreal (Grand Trunk Railroad and other lines to Niagara,) of railroad and steamboat, 436 miles. Total, 727 miles.

This great Mecca of the world's worshippers of landscape beauty, the mighty wonder of Niagara, is on its namesake river, a strait connecting the flood of Lakes Erie and Ontario, and dividing a portion of the State of New York on the west from the Provinces of Canada. The cataracts thus lie within the territory both of Great Britain and of the United States. They are some 20 miles below the entrance of the river, at the north-east extremity of Lake Erie, and about 14 miles above its junction with Lake Ontario.

The waters for which the Niagara is the outlet, cover an area of 150,000 square miles—floods so grand and inexhaustible as to be utterly unconscious of the loss of the *ninety millions of tons* which they pour every hour, through succeeding centuries, over these stupendous precipices.

Fortunately, the most usual approach to Niagara—that by the American shore—is the best, all points considered. " The descent of about 200 feet, by the staircase, brings the traveller directly under the shoulder and edge of the American Fall, the most imposing scene, for a single object, that he will ever have witnessed. The long column of sparkling water seems, as he stands near it, to descend to an immeasurable depth, and the bright seagreen curve above has the appearance of being set into the sky. The tremendous power of the Fall, as well as the height, realizes his utmost expectations. He descends to the water's edge and embarks in a ferry-boat, which tosses like an egg-shell on the heaving and convulsed water, and in a minute or two he finds himself in the face of the vast line of the Falls, and sees with surprise that he has expended his fullest admiration and astonishment upon a mere thread of Niagara—the thousandth part of its wondrous volume and grandeur. From the point where he crosses to Table Rock, the line of the Falls is measurable at three-quarters of a mile; and it is this immense extent which, more than any other feature, takes the traveller by surprise. The tide at the ferry sets very strongly down, and the athletic men who are employed here keep the boat up against it with difficulty. Arrived near the opposite landing, however, there is a slight counter-current, and the large rocks near the shore serve as a breakwater, behind which the boat runs smoothly to her moorings.'*

It is from the American side of the

* This passage is from "American Scenery," and since it was written a fairy little steamer has been employed to traverse the vexed river, and the timid cross readily upon the grand Suspension Bridge.

Niagara Falls. The Rapids.

river that access is had to the hundred points of interest and surprise in the famous Goat Island vicinage, with its connecting bridges, its views of the Rapids, of the Cave of the Winds, of the scene of Sam Patch's great leap, and of its bold over-topping tower; and in other neighborhoods of the Whirlpool, of the Chasm Tower, and the Devil's Hole.

A totally different and not less wonderful gallery of natural master-pieces is opened upon the Canada shore. The terrible marvels of the Table Rock above, and of Termination Rock behind the mighty Horse-Shoe Fall; the noble panorama from the piazzas of the Clifton House, the Burning Spring, the historical village of Chippewa, and the battle field of Lundy's Lane; Bender's Cave, etc.

GUIDE.—*Hotels.*—Upon the American side of the river, the Cataract House and the International Hotel are most excellent homes for the tourist. The Empire, the Niagara, and the St. Lawrence, are also popular resorts. On the Canada side stop at the Clifton.

Goat Island. (American side.)— Leaving the Cataract House, take the first left-hand street, two minutes' walk to the bridge, which leads to the toll-gate on Bath Island. This bridge is itself an object of curious wonder, in its apparently rash and dangerous position. It is, however, perfectly safe, and is crossed hourly by heavily laden carriages.

The Rapids are seen in grand and impressive aspect on the way to Goat Island. The river descends 51 feet in a distance of three-quarters of a mile by this inextricable turmoil of waters. It is one of the most striking incidents in the Niagara scenery. Standing on the bridge, and gazing thence up the angry torrent, the leaping crests seem like " a battle-charge of tempestuous waves animated and infuriated against the sky. The rocks, whose soaring points show above the surface, seem tormented with some supernatural agony, and fling off the wild and hurried waters, as if with the force of a giant's arm. Nearer the plunge of the Fall, the Rapids become still more agitated:

Localities at Niagara.

and it is impossible for the spectator to rid himself of the idea that they are conscious of the abyss to which they are hurrying, and struggle back in the very extremity of horror. This propensity to invest Niagara with a soul and human feelings is a common effect upon the minds of visitors, in every part of its wonderful phenomena. The torture of the Rapids, the clinging curves with which they embrace the small rocky islands that live amid the surge; the sudden calmness at the brow of the cataract, and the infernal writhe and whiteness with which they reappear, powerless from the depths of the abyss—all seem, to the excited imagination of the gazer, like the natural effects of impending ruin—desperate resolution and fearful agony on the minds and frames of mortals." *

Chapin's Island is upon the right of the bridge, within a short distance of the American Fall. It is named in memory of a workman whose life was imperilled by falling into the stream, as he was laboring upon the bridge. Mr. Robinson went gallantly and successfully to his relief in a skiff.

The Toll Gate is upon Bath Island, where baths, warm and otherwise, are accessible at all times to visitors. A fee of 25 cents, paid here, gives you the freedom of Goat Island, during all your stay, be it for the year or less. Near this point are Ship and Big Islands. There is here a very extensive paper-mill.

Another small bridge, and we are upon Iris, or Goat Island. The only place of habitation here is a house at which the traveller can supply himself with refreshments of all inviting kinds, and store his trunks with every variety of samples of Indian ingenuity and labor. The place is called the *Indian Emporium*. Three routes over the island diverge at this point. The principal path followed by most visitors is that to the right, which keeps the best of the sights, as Wisdom always does, until the last; affording less striking

* "American Scenery."

views of the Falls than do the other routes, at first, but far surpassing them both in its grand revealments at the end. This way conducts to the foot of the island, while the left-hand path seeks the head, and the middle winds across. Taking the right-hand path, then, from the Toll Gate, we come, first, to the centre Fall, called **The Cave of the Winds,** mid-distance nearly, between the American and the Horse-Shoe Falls. This wonderful scene is best and most securely enjoyed from the spacious flat rock beneath. The cave is 100 feet high, and of the same extent in width. You can pass safely into the recess behind the water, to a platform beyond. Magical rainbow-pictures are formed at this spot; sometimes bows of entire circles and two or three at once, delight the vision.

At the foot of Goat Island the *Three Profiles* is an object of curious interest. These profiles, seemingly some two feet long, are to be seen, one directly above the other, as you look across the first sheet of water, directly under the lowest point of rock.

Luna Island is reached by a foot bridge, from the right of Goat Island. It has an area of some three-quarters of an acre. The effective rainbow forms seen at this point have given it the name it bears. The venturesome visitor may get some startling peeps far down into the great caldron of waters. A child of eight years once fell into the torrent at this point, and was lost, together with a gallant lad who jumped in to rescue her.

Sam Patch's Leap.—It was upon the west side of Goat Island, near Biddle's Stairs, which we shall next look at, that the immortal jumper, Sam Patch, made two successful leaps into the waters below, saying, as he went off, to the throng of spectators, that "one thing might be done as well as another!" The fellow made his jump too much, within the same year (1829) over the Genesee Falls, at Rochester.

Biddle's Stairs, on the west side of the island, was named after Nicholas Biddle, of United States Bank fame, by whose order they were built. "Make us something," he is reported to have said to the workmen, "by which we may descend and see what is below." At the base of these spiral stairs, which are secured to the rocks by strong iron fastenings, there are two diverging paths. The *up* river way, towards the Horse-Shoe Fall, is difficult, and much obstructed by fallen rocks; but down the current a noble view is gained of the centre Fall or Cave of the Winds. Re-ascending the Biddle Stairs, we come, after a few rods' travel, to a resting place at a little house, and thence we go down the bank, and, crossing a bridge, reach

Prospect Tower.—This precarious-looking edifice, which seems to have "rushed in, as fools do, where angels fear to tread," is very near the edge of the precipice, above which it rises some 45 feet in air. From the top, which is surrounded by an iron railing, a magnificent scene is presented—a panorama of the Niagara wonders—the like of which can be seen from no other point.

The Horse-Shoe Fall, which leads the host of astonishments in this astonishing place, is the connecting link between the scenes of the American and of the Canadian sides of the river, always marvellous from whatever position it is viewed. This mighty cataract is 144 rods across, and, it is said by Prof. Lyell, that fifteen hundred millions of cubic feet of water pass over its ledges every hour. One of the condemned lake ships (the Detroit) was sent over this Fall in 1829, and, though she drew 18 feet of water, she did not touch the rocks in passing over the brink of the precipice, showing a solid body of water, at least some 20 feet deep, to be *above* the ledge. We shall return to the Horse-Shoe Fall from the Canada side.

Gull Island, just above, is an unapproachable spot, upon which it is not likely or possible that man has ever yet stood. There are three other small isles seen from here, called the *Three Sisters.* Near the Three Sisters, on Goat Island, is the spot remembered as the resort of an eccentric, and which is called, after him, the *Bathing Place of Francis Abbott the Hermit.* At the head of Goat Island is Navy Island, near the Canada shore. It was the scene of incidents in the Canadian rebellion of 1837–8, known as the McKenzie War. Chippewa, which held at that period some 5,000 British troops, is upon the Canadian shore below. It was near Fort Schlosser, hard by, that, about this period, the American steamboat Caroline, was set on fire and sent over the Falls, by the order of Col. McNabb, a British officer. Some fragments of the wreck lodged on Gull Island, where they remained until the following spring.

Grand Island, which has an area of 17,240 acres, was the spot upon which Major M. M. Noah hoped to assemble all the Hebrew populations of the world.

Near the Ferry (American side still) there was once an observatory or Pagoda, 100 feet high, from which a grand view of the region was gained. This spot is called *Point View.*

The Whirlpool.—Three miles below the Falls (American side) is the Whirlpool, resembling in its appearance the celebrated Maelstrom on the coast of Norway. It is occasioned by the river making nearly a right angle, while it is here narrower than at any other place, not being more than 30 rods wide, and the current running with such velocity as to rise up in the middle 10 feet above the sides. This has been ascertained by measurement. There is a path leading down the bank to the Whirlpool on both sides, and, though somewhat difficult to descend and ascend, it is accomplished almost every day.

The Devil's Hole is a mile below the Whirlpool. It embraces about two acres, cut out laterally and perpendicularly in the rock by the side of the river, and is 150 feet deep. An angle of this hole or gulf comes within a few feet of the stage-road, affording travellers an opportunity, without alighting, of looking into the yawning abyss. But they should alight, and pass to the further side of the flat projecting rock, where they will feel themselves richly repaid for their trouble.

Chasm Tower, 3¼ miles below the Falls, is 75 feet high, and commands fine views (seen, if you please, of all hues, through a specular medium) of all the country round. A fee of 12½ cents is required.

The Maid of the Mist.—The landing of that singular feature of these wild scenes, the steamboat Maid of the Mist, is two miles below the Falls, whose troublous brink she touches in her frequent trips across the river.

The Great Suspension Bridge spans the chasm at this point. Its total length, from centre to centre of the towers, is 800 feet ; its height above the water, 258 feet. The first bridge, which was built by Mr. Charles Ellett, was a very light and fairy-like affair, in comparison with the present substantial structure. The bridge, as it now stands, was constructed under the directions of Mr. John A. Roebling, at a cost of $500,000. It was first crossed by the locomotive March 8, 1855. Twenty-eight feet below the floor of the railway tracks a carriage and foot-way is suspended. This bridge is used at present by the New York Central, the Erie, and the Great Western, Canada roads.

We will now cross the river on the Suspension Bridge, and explore the wonders of the opposite shores.

Taking a carriage at our hotel, on the American side, we may " do " the Canadian shore very comfortably between breakfast and dinner, if we have no more time to spare. The regular price of carriage hire at the livery stables is one dollar per hour. On the plank road, going and returning, five cents ; at the bridge, for each foot passenger, going and returning the same day, 25 cents, or 12½ each way. If the passenger does not return, the bridge toll is still 25 cents. For each carriage (two horses), going and returning, 50 cents for each passenger, and 50 cents besides for the carriage. A plank road leads from the opposite terminus of the bridge to the Clifton House.

At the bridge is shown a basket in which Mr. Elliott, his wife, and other ladies and gentlemen, crossed over the river on a single wire, about one inch in diameter. A perilous journey across such a gorge and at an elevation in the air of 230 feet! Two or three persons thus crossed at a time, the basket being let down on an inclined plane to the centre of the towers (this was during the building of the first Suspension Bridge), and then drawn up by the help of a windlass to the opposite side. The usual time in crossing was from three to four minutes. By the means of this basket the lives of four men were once saved, when the planks of the Foot Bridge were blown off in a violent storm, and they were suspended over the river by only two strands of wire, which oscillated with immense rapidity, 60 or 70 feet. The basket was sent to their relief, at a moment when the hurricane grew less fearful, and they descended into it by means of a ladder, one at a trip only, until all were released from their terrible position.

Bender's Cave is midway between the Suspension Bridge and the Clifton House. It is a recess six feet high and twenty in length, made by a decomposition of the lime-stone.

If the tourist prefer it, he may cross the river by the ferry, the only route of other days. From the ferry-house the cars descend to the water's edge on an inclined plane of 31 degrees. They are worked by water-power. The time required to make this descent and to cross to the Canada shore is about ten minutes. During the 40 years it is said that this ferry has been in operation, not one life has been lost, nor has any serious accident occurred. We have described the passage of the river in the opening of our article. Upon landing, plenty of carriages will be always found in readiness, as at all other starting and stopping places about the Falls. It will be well to ascertain the fares before employing any of them.

The Clifton House is an old and very favorite resort here, for its home luxuries and for its noble position,

Niagara Falls, from the Clifton House, Canada.

overlooking the river and Falls. It was the residence of Mdlle. Jenny Lind during her visit to Niagara. "The Clifton House," writes Mr. Willis, from whose descriptions of these scenes we have already quoted, "stands nearly opposite the centre of the irregular crescent formed by the Falls; but it is so far back from the line of the arc, that the height and grandeur of the two cataracts, to an eye unacquainted with the scene, are respectively diminished. After once making the tour of the points of view, however, the distance and elevation of the hotel are allowed for by the eye, and the situation seems most advantageous. This is the only house at Niagara where a traveller, on his second visit, would be content to live."

"The *ennui* attendant upon public-houses can never be felt at the Clifton House. The most common mind finds the spectacle, from its balconies, a sufficient and untiring occupation. The loneliness of uninhabited parlors, the discord of baby-thrummed pianos, the dreariness of great staircases, long entries, and bar-rooms filled with strangers, are pains and penalties of travel never felt at Niagara. If there is a vacant half hour to dinner, or if indisposition to sleep create that sickening yearning for society which sometimes comes upon a stranger in a strange land, like the calenture of a fever—the eternal marvel going on without is more engrossing than friend or conversation; more beguiling from sad thoughts than the Corso in carnival time. To lean over the balustrade, and watch the flying of the ferry-boat below, with its terrified freight of adventurers, one moment gliding swiftly down the stream in the round of an eddy, the next lifted up by a boiling wave, as if it were tossed from the scoop of a giant's hand beneath the water; to gaze, hour after hour, into the face of the cataract; to trace the rainbows, delight like a child in the shooting spray-clouds, and calculate fruitlessly and endlessly, by the force, weight, speed, and change of the tremendous waters—is amusement and occupation enough to draw the mind from any thing—to cure madness or create it."

Table Rock.—The grand overhanging platform called Table Rock, and the fearful abysmal scene at the very base of the mighty Horse Shoe Fall, which it presents, is one of the cardinal wonders of Niagara. If one would listen to the terrible noise of the great cataract, let him come here where the sound of its hoarse utterance drowns all lesser sounds, and his own speech is inaudible to himself.

Termination Rock is a recess behind the centre of the Horse Shoe Fall, reached by the descent of a spiral stairway from Table Rock, the traverse for a short distance of the rude marge of the river, and then of a narrow path over a frightful ledge and through the drowning spray, behind the mighty Fall.

Before descending, the visitors make a complete change of toilette, for a rough costume more suitable for the stormy and rather damp journey before them. When fully equipped, their ludicrous appearance excites, for a while, a mirthful feeling, in singular contrast with the solemn sentiment of all the scene around them. This strange expedition, often made even by ladies, has been thus described: "The guide went before, and we followed close under the cliff. A cold, clammy wind blew strong in our faces from the moment we left the shelter of the staircase, and a few steps brought us into a pelting, fine rain, that penetrated every opening of our dresses and made our foothold very slippery and difficult. We were not yet near the sheet of water we were to walk through; but one or two of the party gave out and returned, declaring it was impossible to breathe; and the rest, imitating the guide, bent nearly double to keep the beating spray from their nostrils, and pushed on, with enough to do to keep sight of his heels. We arrived near the difficult point of our progress; and in the midst of a confusion of blinding gusts, half deafened, and more than half drowned, the guide stopped to give us a hold of his skirts

and a little counsel. All that could be heard amid the thunder of the cataract beside us was an injunction to push on when it got to the worst, as it was shorter to get beyond the sheet than to go back; and with this pleasant statement of our dilemma, we faced about with the longest breath we could draw, and encountered the enemy. It may be supposed that every person who has been dragged through the column of water which obstructs the entrance to the cavern behind this cataract, has a very tolerable idea of the pains of drowning. What is wanting in the density of the element is more than made up by the force of the contending winds, which rush into the mouth, eyes and nostrils, as if flying from a water-fiend. The "courage of worse behind" alone persuades the gasping sufferer to take one desperate step more.

It is difficult enough to breathe within; but with a little self-control and management, the nostrils may be guarded from the watery particles in the atmosphere, and then an impression is made upon the mind by the extraordinary pavilion above and around, which never loses its vividness. The natural bend of the falling cataract, and the backward shelve of the precipice, form an immense area like the interior of a tent, but so pervaded by discharges of mist and spray, that it is impossible to see far inward. Outward the light struggles brokenly through the crystal wall of the cataract; and when the sun shines directly on its face, it is a scene of unimaginable glory. The footing is rather unsteadfast, a small shelf composed of loose and slippery stones; and the abyss below boils like—it is difficult to find a comparison. On the whole, this undertaking is rather pleasanter to remember than to achieve.

The *Museum*, near Table Rock, contains more than 10,000 specimens of minerals, birds, fish and animals, many of which were collected in the neighborhood of the Falls. Admittance, 25 cents. The Burning Spring is near the water, two miles above the Falls. The carbonated sulphuretted hydrogen gas here, gives out a brilliant flame when lighted. Charge, 12½ cents.

The height of the Falls is 165 feet. The roar of the waters has been heard at Toronto, 44 miles away, and yet in some states of wind and atmosphere, it is scarcely perceptible in the immediate neighborhood. Niagara presents a new and most unique aspect in winter, when huge icicles hang from the precipices, and immense frozen piles of a thousand fantastic shapes glitter in the bright sun light. Father Hennepin, a Jesuit missionary, was the first European who ever saw Niagara. His visit was in 1678. Niagara is an Indian word of the Iroquois tongue, from Ongakarra, meaning mighty or thundering water.

In the vicinity of Niagara is Lewistown, seven miles distant, at the head of navigation on Lake Ontario—and directly opposite Lewistown is Queenstown, under Queenstown Heights—a famous battle-ground. Brock's Monument, a column of 126 feet, crowns the Heights.

THE ADIRONDACK MOUNTAINS—THE SARANAC LAKES, ETC.

The upper part of the State of New York, lying west and south of Lake Champlain and the St. Lawrence River, respectively, is still a wild primitive forest region, of the highest interest to the tourist, for its wonderful natural beauties and for the ample facilities it offers for the pleasures of the rod and the rifle. Fine mountain peaks stud the whole region, and charming lakes and lakelets are so abundant that travel here is made by water instead of by land—traversing the ponds in row-boats which are carried by easy portage from one lovely brook or lake to another.— Deer fill the woods, and trout are unsuspecting in the transparent floods everywhere. This wilderness land is visited at various points under distinctive names, as the hunting-grounds of the Saranacs, of the Chateaugay woods, of the Adirondacks, and of Lake Pleasant, etc. We shall speak of these several divisions, briefly, in order.

The Saranac Lakes. These wonderful links of the great chain of mountain waters in upper New York, are about a dozen in number, large and small. They lie principally in Franklin County, and may be most readily reached by stage from Wesport or from Keeseville, about midway on the western shore of Lake Champlain—taking stage or private conveyance thence (30 miles) to the banks of the Lower Saranac—which is the outer edge of civilization in this direc-

Lower Saranac Lake, N. Y.

tion. There is a little village and an inn or two at this point, and here guides and boats, with all proper camp-equipage for forest-life, may be procured. For this route the tourist must engage a boatman, who for a compensation of two or three dollars per day—the price will be no more if he should have extra passengers—will provide a boat, with tent and kitchen apparatus, dogs, rifles, etc. The tourist will supply, before starting, such stores as coffee, tea, biscuit, etc., and the sport by the way, conducted by himself or by his guide, will keep him furnished with trout and venison. If camp life should not please him, he may, with some little inconvenience, so measure and direct his movements as to sleep in some one or other of the shanties of the hunters or of the lumber-men found here and there on the way. The tent in the forest, however, is much preferable.

Leaving the Lower Saranac, we will pass pleasantly along some half a dozen miles—then make a short portage, the guide carrying the huge boat, by a yoke on the back, to the Middle Saranac —there he may go on to the Upper Lake of the same name, and thence by a long portage of three miles to Lake St. Regis. These are all large and beautiful waters, full of delicious islands and hemmed in upon all sides by fine mountain ranges. Trout may be taken readily at the inlets of all the brooks, and deer may be found in the forests almost at will.

Returning from St. Regis, and back via the Upper to the Middle Saranac, we continue our journey, by portage, to the Stony Creek ponds—thence three miles by Stony Creek to the Rackett River—a rapid stream, with wonderful forest vegetation upon its banks. This water followed for some 20 miles, brings us to Tupper's Lake—the finest part of the Saranac region. Tupper's Lake is the largest of this chain, being seven miles long and from one to two miles broad. The shores and headlands and islands are especially bold and picturesque, and at this point the deer is much more easily found than elsewhere in the

neighborhood. Below Tupper's Lake —the waters commingling—is Loughncah, another charming pond. The chain continues on yet for miles, but the Saranac trip, proper, ends here. This mountain voyage and the return to Lake Champlain might be made in a week, but two or three, or even more, should be given to it. It is seldom that ladies make the excursion, but they might do so with great delight. The boatmen and hunters of the region are fine, hearty, intelligent and obliging fellows. That wonderful ravine, the "Walled Banks of the Ausable," (see Index,) should be seen by the Saranac tourist, on his way from Lake Champlain to Keeseville.

The Adirondack Mountains. The Adirondack region may be reached by private conveyance (only) over a rude mountain road from Schroon Lake, above Lake George, or more conveniently from Crown Point village, just beyond the ruins of Fort Ticonderoga, on Lake Champlain. The distance thence is some 30 miles, and requires a day to travel. The tourist in this region will move about by land more than by water, as among the Saranacs; for, although the lakes are numerous enough, it is among and upon the hills that the chief attractions are to be found. The accommodations are rude enough—the only inn being the boarding-house at the village of the Adirondack Iron Works. Stopping at this point, as headquarters, he may make a pleasant journey down Lake Sanford near by, on one side, and upon Lake Henderson on the other hand. In one water he ought to troll for pickerel, and in the other, cast his fly for trout; and upon both, enjoy the noble glimpses of the famous mountain peaks of the Adirondack group, the cliffs of the Great Indian Pass, of Mounts Colden, M'Intyre, Echo Mountain, and other bold scenes. It will be a day's jaunt for him afterwards to explore the wild gorge of the Indian Pass, five miles distant; another day's work to visit the dark and weird waters of Avalanche Lake; and yet another to reach the Preston Ponds, five miles in a

8

The Adirondacks—Lake Pleasant.

Lake Henderson, in the Adirondacks, N. Y.

different direction. He will find, indeed, occupation enough for many days, in exploring these and many other points, which we may not tarry to catalogue; and, in any case, he must have 48 hours to do the tramp, *par excellence*, of the Adirondack—the visit to the summit of the brave Tehawus, or Mount Marcy, the monarch of the region. Tehawus is 12 miles away, and the ascent is extremely toilsome.

The Adirondacks (named after the Indian nation which once inhabited these fastnesses) lie chiefly in the county of Essex, though they extend into all the jurisdictions around. Mount Marcy, or Tehawus, "the Cloud Splitter," is 5,467 feet high. Mount M'Intyre has an elevation almost as great. The Dial Mountain, M'Marten, and Colden are also very lofty peaks, impressively seen from the distance, and inexhaustible in the attractions which their ravines, and brooks, and waterfalls present. White Face and other grand hill peaks belong to the neighboring range called the Keene Mountains. The Hudson River rises in this wilderness.

Lake Pleasant. To reach Lake Pleasant, and the adjoining waters of Round, Piseco, and Louis Lakes—a favorite and enchanting summer resort and sporting ground—take the Central Rail-

way from Albany, as far as Amsterdam, and thence, by stage or carriage, to Holmes' Hotel, on Lake Pleasant. The ride from Amsterdam is about 30 miles. The stage stops overnight at a village, *en route*. Mr. Holmes' house is an excellent place, with no absurd luxuries, but with every comfort for which the true sportsman can wish. It is a delightful summer home for the student, and may be visited very satisfactorily by ladies. The wild lands and waters here are a part of the lake region of northern New York, of which we have already seen something on the Saranacs, and among the Adirondacks. The Saranac region is connected with Lake Pleasant by intermediate waters and portages. The deer, and other game, is abundant here in the forests, and fine trout may be taken in all the brooks and lakes. Lake Pleasant and its picturesque *confrères*, lie in Hamilton County.

All this northern part of New York is quite similar in its attractions to the wilderness in the upper part of the State of Maine.

TO LAKE MAHOPAC.

From New York, via Harlem Railroad the depot up town, at the corner of Fourth avenue and Twenty-sixth street,

New York, 51 miles, to Croton Falls Station. (See Harlem Route from New York to Albany.)

Stages leave Croton Falls for Lake Mahopac, five miles, on the arrival of the cars; stage fare, 25 cents.

Lake Mahopac, a favorite summer resort, in the immediate vicinity of New York, and much frequented by its citizens, both for a day's excursion and as a continued home, lies in the western part of the town of Carmel, Putnam County, New York, 13 miles east from Peekskill, on the Hudson, and five miles from Croton Falls Station, on the Harlem Railroad. The lake is nine miles in circumference, and is about 1800 feet above the sea. It is one of the principal sources of supply to the Croton. Though the landscape has no very bold features, but little to detain the *artist*, yet its quiet waters, its pretty wooded islands, the romantic resorts in its vicinage, the throngs of pleasure-seeking strangers, the boating, and fishing, and other rural sports, make it a most agreeable spot for either a brief visit or long residence. There are many attractive localities of hill and water scenery around Mahopac. There are pleasant hotels here, well filled during the season by boarders or by passing guests. It is a nice retreat to those whose business in the great city below forbids their wandering far away.

LEBANON SPRINGS AND SHAKER VILLAGE.

Shaker Village, at New Lebanon, New York, two miles from the Springs, is a station on the Western Railway, from Boston to Albany—46 miles from Albany.

It may be readily reached from New York by the Hudson River Route to the City of Hudson, and thence by the Hudson and Berkshire, and the Albany and Boston Roads, or by the Harlem Railroad and its connections with the above-mentioned routes.

There are ample accommodations for the traveller at this favorite watering-place, in a well-appointed hotel, a water-cure establishment, &c., pleasantly perched on a hill slope, overlooking a beautiful valley. There are pleasant drives all around, over good roads, to happy villages, smiling lakelets, and inviting spots of many characters. Trout, too, may be taken in the neighborhood. The water of the Springs flows from a cavity 10 feet in diameter, and in sufficient volume to work a mill. Its temperature is 72°. It is soft and pleasantly suited for bathing uses, is quite tasteless and inodorous. For cutaneous affections, rheumatism, nervous debility, liver complaint, &c., it is an admirable remedial agent.

The village of New Lebanon, or the celebrated Shaker settlement, is two miles from the Springs, and is a point of great interest to the visitors there, especially on Sunday, when their singular forms of worship may be witnessed.

SHARON SPRINGS.

From Albany, New York, by the Central Railroad for Buffalo, as far as Palatine Bridge, 55 miles; thence by stage, 10 miles, over a plank road. The most ancient, and perhaps the best hotel is the Pavilion, an establishment large enough to accommodate 300 guests. The Eldridge House, also well-appointed, is nearer to the Springs, and is less expensive than the Pavilion. The scenery of the neighborhood is highly attractive, and the panorama, commanded by the eminence on which the Pavilion stands, is remarkably fine.

The waters are pure and clear, and although they flow for one-fourth of a mile from their source with other currents, they yet preserve their own distinct character. The fall here is of sufficient force and volume to turn a mill. It tumbles over a ledge of perpendicular rocks, with a descent of some 65 feet. The magnesia and the sulphur springs much resemble the White Sulphur of Virginia.

Cherry Valley is in the vicinage of Sharon Springs, accessible also from Palatine Bridge, on the Albany and Buffalo road, and from Canajoharie, on the Erie Canal, from which it lies about 26 miles in a south-west direction.

Otsego Lake and Cooperstown, famous as the home of the late Fenimore Cooper the novelist, are near by.

COLUMBIA SPRINGS.

From New York, by Hudson River Railway, or steamboats to Hudson; thence, by carriage or stage, four miles.

The Columbia Springs have of late years grown into great popular favor. They are easily accessible, lying only four miles from the City of Hudson. They are within the town of Stockport, Columbia County, New York. The site and grounds are highly varied and picturesque, jumping delightfully from hill to dale, from forest glen to grassy lawn.

There is, too, a merry brooklet, which winds coquettishly through the landscape, affording now a quiet slope for some "melancholy Jacques," now a dashing cascade for him of brighter mood. In the immediate neighborhood, moreover, there flows a larger water, offering all the country charms of boating and fishing. The hotel here is large and well appointed, and Mr. Charles B. Nash, the enterprising proprietor, is every year swelling its conveniences and comforts, and adding to the seductions of the occupations and enjoyments, and to the beauty of the scenery out of doors.

AVON SPRINGS.

The Avon Springs may be reached by the Central Railway from Albany to Buffalo, via Rochester, from which city they are distant 20 miles. The village of Avon is upon the Genesee River, which it overlooks from a charming terrace 100 feet above. On this lofty position the picturesque landscape of the neighborhood is seen to great advantage. The Springs are near at hand, a little to the south-west. With ample hotel conveniences and enjoyments, the Avon Springs meet the popular favor they so well deserve.

RICHFIELD SPRINGS, N. Y.

Richfield Springs are reached from Herkimer, 81 miles from Albany, on the Central Road to Buffalo. They are in the town of Richfield, Otsego County, south-east of Utica, near the head of Canaderaga, one of the numerous lakes of all this part of New York. Otsego Lake is six miles distant; and another six miles will take the traveller to Cooperstown. Cherry Valley, Springfield, and other villages are near by.

LONG ISLAND.

Long Island, a part of the State of New York, is 115 miles in length, and, at some points, about 20 in breadth; with the Atlantic on the south, and the Long Island Sound on the north. The upper part of the island is agreeably diversified with hills, though the surface is for the most part strikingly level. The coast is charmingly indented with bays; and delicious fresh-water ponds, fed by springs, are everywhere found on terraces of varying elevation. These little lakes, and the varied coast-views give Long Island picturesque features, which, if not grand, are certainly of most attractive and winning character, yet heightened by the rural beauty of the numerous quiet little towns, and charming summer villas.

The places in the immediate vicinity of New York, we have already mentioned among the suburban resorts of that city, and we might almost have included the whole island in that classification, so easily is every part reached either by the steamboats, or by the railway which traverses the length of the island, from Brooklyn 95 miles to Greenport.

The lower shore of the island, which is a net-work of shallow, land-locked waters, extending 70 miles, is the resort of innumerable flocks of aquatic fowl, and thither go the New York sportsmen or gunners for pastime, and glory for themselves, and for delights for the tables of their city friends. In no other part of the Union is there a greater variety and abundance of wild birds than on this coast, and no where else are they more systematically sought. To answer the wants of the sportsmen,

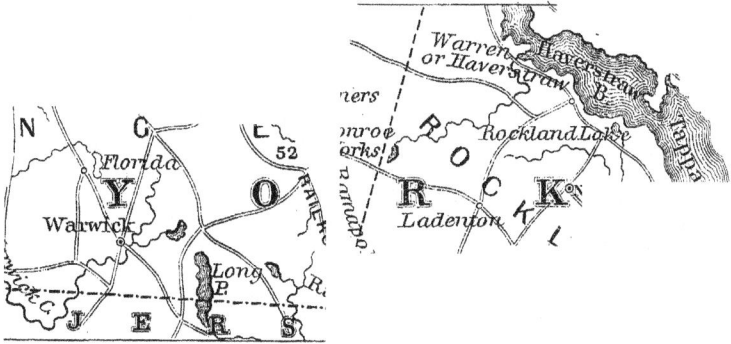

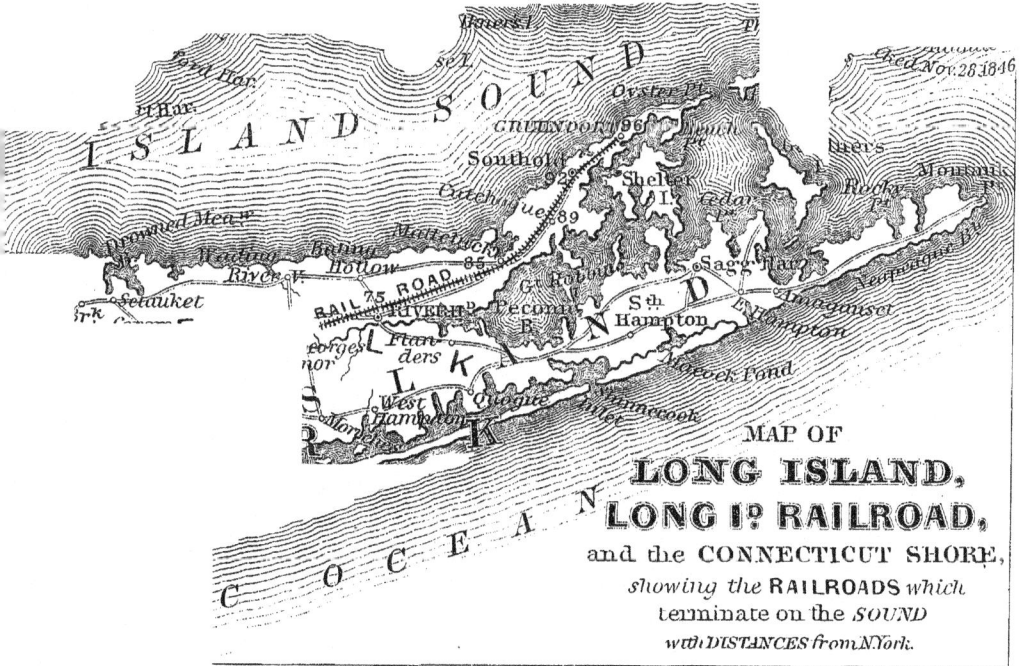

MAP OF
LONG ISLAND,
LONG I? RAILROAD,
and the **CONNECTICUT SHORE,**
showing the **RAILROADS** *which*
terminate on the *SOUND*
with DISTANCES from N.York.

excellent accommodations have been every where provided in the way of comfortable hotels and boat equipage.

Cedarmere, the home of the Poet Bryant, is near the pretty village of Roslyn, at the head of Hempstead Bay, about two hours' journey from New York by steamboat to Glen Cove, and thence by stage ; or by the Long Island Railway 20 miles to Hempstead Branch, and by connecting stages. Cedarmere is a spot of great, though quiet picturesque beauty, overlooking Hempstead Bay, and the Connecticut shore across the Sound. Many of the charming, terraced, spring-water lakes of which we have spoken already, as among the pleasant and unique features of the Long Island landscape, are found within the domain of Cedarmere, in the village of Roslyn, and, indeed, through all the vicinage for miles around. Within a pleasant stroll of Mr. Bryant's residence is Hampstead Hill, the highest land on Long Island. This fine eminence overlooks the Sound and its inlets on the one hand, and the ocean beach on the other ; at its base the village of Roslyn is nestled among green trees, and placid lakelets. The house at Cedarmere makes no architectural pretensions, though it falls most agreeably into all the charming pictures, which every changing step over the hills, or along the margin of the ponds presents to view.

Battle of Long Island (August 1776). The thoughts of the tourist on the quiet pastoral plains of Long Island, will revert with interest to that eventful night when the British troops under Sir Henry Clinton, Lord Cornwallis, and General Howe, made their silent, unsuspected march from Flatlands, through the swamps and passes to Bedford Hills, stealing upon the rear, and almost surrounding the patriot lines ; " that able and fatal scheme which cost the Americans the deadly battle of Long Island, with the loss of nearly 2,000 out of the 5,000 men engaged."

The surprise of the attack, the obstinacy of the conflict, the bold retreat, and the loss of the city of New York, to which it led, make this battle one of the most romantic episodes in the history of the Revolution.

"Never," says Mr. Irving, " did retreat require greater secrecy and circumspection. Nine thousand men, with all the munitions of war, were to be withdrawn from before a victorious army, encamped so near that every stroke of the spade and pick-axe from their trenches could be heard.

" The retreating troops, moreover, were to be embarked and conveyed across a strait, three-quarters of a mile wide, swept by rapid tides. What with the greatness of the stake, the darkness of the night, the uncertainty of the design, and the extreme hazard of the issue, it would be difficult to conceive a more deeply solemn and interesting scene.

"Washington wrung his hands in agony at the sight of this fatal battle. 'Good God!' cried he, as his troops were swept down, ' what brave fellows I must lose to-day!'"

NEW JERSEY.

SETTLEMENTS were made in this State at Bergen, by the Dutch, soon after their arrival in New York. In 1627 a Swedish colony was founded near the shores of the Delaware, in the south-western part of the State. A droll account of the quarrels of these Swedish folk with the Dutchmen of New Amsterdam, may be found in Diedrick Knickerbocker's solemn history of the Amsterdam colonists. New Jersey is one of the old Thirteen States. She did her part nobly in the long war of Independence. The famous battles of Trenton, and of Princeton, and of Monmouth, at all of which Washington was present and victorious, occurred within her limits. Morristown was the winter camp of the American army in 1776–'77.

New Jersey has not a very wide territory, yet she presents many natural attractions to the traveller. Her sea-coast abounds in favorite bathing and sporting resorts, much visited by the citizens of New York and Philadelphia. Among these Summer haunts are Cape May, Long Branch, Sandy Hook, Absecum Beach, Deal, Squam Beach, and Tuckerton.

In the southern and central portions of this State the country is flat and sandy; in the north, are some ranges of picturesque hills, interspersed with charming lakes and ponds. Some of the Alleghany ridges traverse New Jersey, forming the spurs known as Schooley's Mountain, Trowbridge, the Ramapo, and Second Mountains. In the north-western part of the State, are the Blue Mountains. The Nevisink Hills rising nearly 400 feet on the Atlantic side, are usually the first and last land seen by ocean voyagers as they approach and leave New York. The celebrated Palisade Rocks of the Hudson River are in this State.

NEW YORK TO PHILADELPHIA.

There are two great routes between the cities of New York and Philadelphia, one known as the New Jersey railway line, and the other as the Camden and Amboy route. The former is the most expeditious; the latter, being partly by water, is the most agreeable in summer time.

The New Jersey Railway Route.

This route passes over the New Jersey, and the Philadelphia, and Trenton Railroads. Leaves New York at foot of Cortland street (by ferry across the Hudson to Jersey City), several times each day. Distance, 87 miles; time (express trains), four hours.

STATIONS.

NEW YORK—Jersey City, 1 mile; Newark, 9; Elizabeth, 15; Rahway, 19; Uniontown, 23; Metuchin, 27; NEW BRUNSWICK, 31; Dean's Pond, 39; Kingston, 43; PRINCETON, 47; Trenton, 57; Bordentown, 62; Burlington, 68; CAMDEN, 86; Bristol, 70; Cornwell's, 74; Tacony, 79; Kensington, 85; PHILADELPHIA, 87.

This route, lying as it does, between the two greatest cities on the continent of America, is an immense thoroughfare, over which floods of travel pour unceasingly by day and by night. The region is populous and opulent, and necessarily thronged with towns and villages, and villas; for 20 or 25 miles from each terminus, over which the two cities spread their suburbs, the crowded trains are passing and repassing momently.

Leaving Jersey City (see New York and vicinity), the track over which we pass for two miles, is that used also by the great Erie Railway, and which is traversed by all the thousands daily voyaging from every part of the Canadas, the New England States, and New York, for any and all regions of the wide South. Perhaps no other two miles of railway in the world bears such prodigious freights of men and merchandise as this.

Newark, 9 miles from New York, and 78 from Philadelphia; settled in 1666; population (in 1855) 50,000, is upon the right bank of the Passaic River, 4 miles from its entrance into Newark Bay. It is built on an elevated plain, regularly laid out in wide streets, crossing at right angles. Many portions of the city are elegantly built, and in its most recherché quarter, are two charming parks, filled with noble elms. Among its most imposing public edifices, are the Court House, an Egyptian structure of brown stone; the Market street Railway Depôt; and the Library Buildings, also in Market street.

Among the chief literary institutions of the city are, the Library Association, the State Historical Society, the New Jersey Literary Association, the Wesleyan Institute, and the Association of Teachers and Friends of Education. Among its forty churches are many

MAP OF THE
G.t TRAVELLING ROUTES
between
N. YORK & PHILADE
WITH OTHER RAILROADS
and
PLACES OF RESORT.

Newark, New Jersey.

very imposing edifices. Of these, particular mention may be made of the Catholic, on Washington street; the new Presbyterian churches—one near the Lower Park, and the other in High street; and the Methodist in Market street. Other elegant church structures are now in progress. In each of the seven wards of the city there is a fine public school, with departments for both sexes, and a total number of pupils of 2,500.

Newark is distinguished for its manufactures, which are large and prosperous. Steamboats, as well as railways, connect it with New York. It is the eastern terminus of the Morris and Essex Railroad and of the Morris Canal.

Elizabeth.—Population at this time, about 10,000; 15 miles from New York; is situated upon the Elizabethtown Creek, two miles from its entrance into Staten Island Sound. It was once the capital and chief town of the State. Here diverges the N. J. Central R. R., 61 miles hence, into the interior, at Easton, Penn., on the Delaware river.

Rahway, N. J. — Population (in 1853), about 6,000; 19 miles from New York; lies upon both sides of the Rahway River. The Rahway Female Institute is a flourishing establishment,

and so the Union School and a Classical Boarding-school for boys. Rahway is noted for its manufactures of carriages, stoves, hats, earthenware, etc. Some 3,000 vehicles are annually sent hence to the Southern markets.

New Brunswick.—Population (in 1850), 18,000, is at the head of steamboat navigation on the Raritan River. This is the seat of Rutgers College and School, founded in 1770. The streets on the river are narrow and crooked, and the ground low; but those on the upper bank are wide, and many of the dwellings are very neat and elegant, surrounded by fine gardens. From the site of Rutgers College on the hill, there is a wide prospect, terminated by mountains on the north, and by Raritan Bay on the east. The Delaware and Raritan Canal extends from New Brunswick to Bordentown, on the Delaware River, 42 miles. This canal is 75 feet wide and 7 feet deep, and is navigated by sloops and steamboats of 150 tons. This fine work cost $2,500,000. The railway here crosses the Raritan River.

Princeton.—Population, 2,500; built on an elevated ridge; is a pleasant little town, of literary and historical interest. It is the seat of the Princeton College, one of the oldest and most famous educational establishments in

New York to Philadelphia by the Camden and Amboy Route.

the country, founded by the Presbyterians at Elizabethtown, 1746, and removed to Princeton in 1757. Here, also, is the Theological Seminary of the Presbyterian church, founded 1812. In this vicinity was fought the memorable battle of January 3, 1777, between the American forces under General Washington, and those of the British, under Lieutenant-Colonel Mawhood, in which the latter were vanquished.

Trenton, the capital of New Jersey; population (in 1850), 6,460; is on the left bank of the Delaware, 30 miles from Philadelphia, and 57 from New York. The city is regularly laid out, and has many fine stores and handsome dwellings. The State House, which is 100 feet long and 60 wide, is built of stone, and stuccoed, so as to resemble granite. Its situation, on the Delaware, is very beautiful, commanding a fine view of the river and vicinity. Here is the State Lunatic Asylum, founded in 1848, and also the State Penitentiary. Trenton has two daily and two other newspapers, 17 churches, and a State Library. The city is lighted with gas. The Delaware is here crossed by a handsome covered bridge, 1,100 feet long, resting on five arches, supported on stone piers, and which is considered a fine specimen of its kind. It has two carriage-ways, one of which is used by the railroad. The Delaware and Raritan Canal, forming an inland navigation from New Brunswick, passes through Trenton to the Delaware at Bordentown. It is supplied by a navigable feeder, taken from the Delaware 23 miles north of Trenton. It was completed in 1834, at a cost of $2,500,000. The Delaware and Raritan Canal passes through the city, and connects it with New York and Philadelphia. At this point the New Jersey Railroad, which we have thus far travelled, 57 miles from New York, ends, and the Philadelphia and Trenton, upon which we make the rest of our journey, begins. A branch road, six miles long, connects with the Camden and Amboy Railway at Bordentown. The Belvidere, Delaware, and Flemington Railroad leads

hence, 63 miles, to Belvidere, in the interior, along the Delaware River.

Here was fought the famous

Battle of Trenton.—On Christmas night, in 1776, and during the most gloomy period of the Revolutionary war, General Washington crossed the Delaware with 2,500 men, and early on the morning of the 26th commenced an attack upon Trenton, then in possession of the British. So sudden and unexpected was the assault, that of the 1,500 German troops encamped there, 906 were made prisoners. This successful enterprise revived the spirit of the nation, as it was the first victory gained over the German mercenaries. General Mercer, a brave American officer, was mortally wounded in the attack.

It was here, upon Trenton Bridge, that occurred the memorable and beautiful reception of Washington, while on his way from New York to Mount Vernon, twelve years after the glorious victory.

Trenton was settled about 1680, and was named in 1720, in honor of Col. Wm. Trent, Speaker of the House of Assembly. [Here we take the Branch road, six miles to Bordentown, and thence by Camden and Amboy line, or continue, as we now shall, by Philadelphia and Trenton route.]

Bristol is a beautiful village, on the west bank of the Delaware, nearly opposite Burlington. The Delaware division of the Pennsylvania Canal, which communicates with the Lehigh at Easton, terminates here in a spacious basin on the Delaware. Pop., 2,570.

Tacony and *Kensington* are within the corporate limits of Philadelphia, at which city we have now arrived. See description of Philadelphia for hotels. We will now follow the line of the second, or Camden and Amboy route.

CAMDEN & AMBOY (OR STEAMBOAT) ROUTE.

FROM NEW YORK TO PHILADELPHIA.

Steamboat for Philadelphia, via Camden and Amboy route, leave Pier No. 1 North River, New York, daily (Sundays

New York to Philadelphia by the Camden and Amboy Route.

excepted), at 6 A.M. and 2 P.M., for South Amboy, 27 miles, and thence by rail. Fare by morning line is $2 25; by the afternoon (*Express*) line, $3.

Camden and Amboy Railroad from South Amboy.—STATIONS : *New York*, South Amboy, 27 miles; South River, —; Spotswood, 38; Jamesburg, 42; Prospect Plains, —; Cranberry Station, 45; Hightstown, 49; Centreville, 53; Newtown, 56; Sandhills, 58; *Bordentown*, 63; Hammel's Turn, 68; Burlington, 71; Beverly, 77; Rancocas, 78; Palmyra, 83; Fish House, 85; *Camden*, 89; *Philadelphia*, 90.

In the summer season, no more delightful journey can be made than the first twenty-seven miles of our present route across the lovely Bay and Harbor of New York, to South Amboy, past the villaed and villaged shores of Staten Island, and the Raritan River. The scenery of this region is described in our chapter upon New York City and its vicinity.

South Amboy is the landing place, and also the terminus of the Camden and Amboy Railroad. Upon our arrival there, we are transported, in a short space of time, from the steamboat to the railroad cars; and after a slight detention, we proceed on our journey up the steep ascent from the river, and

soon enter a line of deep cutting through the sandhills. The road is then continued through a barren and uninteresting region of country, towards the Delaware at

Bordentown, 35 miles from Amboy. Here are the extensive grounds and mansion formerly occupied by the late Joseph Bonaparte, ex-King of Spain, which are among the most conspicuous objects of the place.

Bordentown is situated on a steep sandbank, on the east side of the Delaware. Although in a commanding situation, the view is greatly obstructed from the river. This is a favorite resort of the Philadelphians during the summer season. The Delaware and Raritan Canal here connects with Delaware River. A branch road, running along the canal and river, unites this town with Trenton. Population, 3,000.

Burlington, settled in 1670, and with a present population of about 5,000, is a port of entry on the Delaware, 19 miles from Philadelphia. Burlington College, founded by the Episcopalians in 1846, is located here, and there are besides, upon the banks of the river, two large boarding-schools, one for each sex. Burlington is connected with Philadelphia by steamboat, and is a place of great summer resort thence.

The Passaic Falls, New Jersey.

8*

Greenwood Lake, N. J.

Beverly, built on the banks of the Delaware since 1848, has now a population of from 1,000 to 1,500. It is a suburb of Philadelphia, distant thirteen miles.

Camden is at the terminus of our route, upon the banks of the Delaware River, immediately opposite the city of Philadelphia, to which we now cross by ferry. For further mention of Camden, and for hotels, etc., see description of Philadelphia.

The Falls of the Passaic occur in the town of Paterson, 16 miles from New York, on the route of the Erie Railway. This bold passage on the Passaic, though it has of late years lost much of its ancient beauty, is still a scene of great attraction, particularly when the stream chances to be generously swollen after heavy rains. Paterson itself is an agreeable town of very considerable importance. It has a population of some 20,000.

Cape May.—This fashionable summer resort is at the extreme southern point of New Jersey, where the Delaware Bay enters the sea. The hotel and bathing accommodations here are ample and excellent. Cape May may be reached by steamboats, through the season, from New York and Philadelphia.

SCHOOLEY'S MOUNTAIN, N. J.—BUDD'S LAKE.

Schooley's Mountain, in New Jersey, is a convenient and popular summer resort and residence of the citizens of New York. It is readily reached from New York by the N. J. Central R.R. to White House, or by the Morris and Essex Railway for Hacketstown, and by stage direct to the Mountain. Southerners proceed via Philadelphia and New Brunswick, connecting with the New Jersey Central Railway at Bound Brook, and from this line as above.

The height of the mountain is about 1,100 feet above the sea. The spring is near the summit. It contains muriate of soda, of lime, and of magnesia, sulphate of lime, carbonate of magnesia, and silex and carbonated oxide of iron. The principal hotel is the Belmont Hall.

Budd's Lake, a mountain water, deep and pure, and supplied with fish, is seven miles from the spring, and may also be reached direct by railway.

GREENWOOD LAKE.

To Sloatsburg, 35 miles from New York, by the N. Y. and Erie Railway, and thence 12 miles (tri-weekly) by stage; or from Newburgh on the Hudson to Chester Station, Erie route, and from there eight miles by stage.

To Greenwood Lake, sometimes called Long Pond, is a very agreeable jaunt from the metropolis, whether for the pure air of the hills, the pleasant aspects of nature, or for the sports of the angle and the gun. Greenwood lies half in New York and half in New Jersey, in the midst of a very picturesque mountain region. It is a beautiful water of seven miles in extent, and all about it, in every direction, are lesser, but scarcely less charming, lakes and lakelets, some of which, as you ride or ramble over the country, delight your surprised eyes, where least dreamed of. Such an unexpected vision is Lake Macopin and the larger waters of Wawayandah.— This last mentioned lake is situated on the Wawayandah Mountains, about 3½ miles from the New York and New Jersey boundary line. The word Wawayandah signifies winding stream, and is very characteristic of the serpentine course of the outlet of this lake towards the Wallkill. Wawayandah is almost divided by an island into two ponds, and thus gets its *home* name of "Double Pond." It is very deep and abounds in fine trout. This varied hill and lake neighborhood presents in its general air an admirable blending of the wild ruggedness of the great mountain ranges and the pastoral sweetness of the fertile valley lands; for it possesses the features of both, though of neither in the highest degree.

There is a comfortable summer hotel at the head of Greenwood Lake. An excursion thither from New York would require two or three days' time.

DELAWARE.

DELAWARE is, after Rhode Island, the smallest State in the Union—her greatest length and breadth being, respectively, only 96 and 37 miles. The first settlements here were made by the Swedes and Finns, about the year 1627. In 1655, the country fell into the possession of the Dutch, and in 1664 passed under British rule. It was originally a portion of Pennsylvania, and was governed by the rulers of that Colony, until the time of the Revolution.

The landscape of the northern portion of Delaware is agreeably varied with modest hills and pleasant vales. In the central and southern portions of the State the country is level, ending in marsh and swamp lands. The only considerable waters are the Delaware River and Bay, on the eastern boundary. The Brandywine is a romantic stream, famous for the Revolutionary battle fought upon its banks near the limits of this State, September, 1777. Lords Cornwallis and Howe, Generals Washington, Lafayette, Greene, Wayne, and other distinguished English and American leaders took part in this memorable conflict. The Americans retreated to Germantown with a loss of 1,200 men, while the British remained in possession of the field, with a loss of about 800.

BALTIMORE—FROM PHILADELPHIA.

The direct routes between Philadelphia and Baltimore, are: the *Railroad Line,* and the *Steamboat* and *Railroad* alternately. The distance by the former route is 97 miles. *Fare,* $3.00: time, *six* hours. By the latter route the distance is 117 miles. *Fare,* $3.00: time, *seven* hours.

There is another railroad route be-

180   DELAWARE.

Route from New York to Philadelphia, via Wilmington and Delaware.

tween the two cities, but it is never passed over by travellers wishing to go direct between Philadelphia, Baltimore and the South. Tourists, whose time is not limited, and who are desirous of varying the route of travel, will find that over the Columbia Railroad to the Susquehanna River, thence to York, and from thence to Baltimore, a very pleasant excursion. Distance, 153 miles. *Fare*, $5.00. Another route open to *tourists*, is from Philadelphia by steamboat down the Delaware River to Delaware City, 44 miles; thence through the Chesapeake and Delaware Canal, 16 miles; and thence down Elk River and Chesapeake Bay and up the Patapsco River to Baltimore, 56 miles: total, 116 miles. The great feature of interest here, and perhaps the only inducement to deviate from the regularly travelled routes, would be the pleasure of seeing the formidable excavation on the Chesapeake and Delaware Canal, termed the "Deep Cut," which extends for *six* miles, and is 70 feet deep in the deepest part. A bridge of 235 feet span extends over this great chasm, at an elevation of 90 feet above the canal, under which steamboats, schooners, and other small vessels can pass. This canal is 66 feet wide at the surface, and 10 feet deep, with two lift and two tide locks, 100 feet long by 22 wide. It was completed in 1829, at a cost of $2,750,000.

PHILADELPHIA, WILMINGTON AND BALTIMORE RAILROAD.—This Road extends from Philadelphia to Baltimore, 97 miles. It is the great thoroughfare between the two cities, and during the winter months, the only travelled route.

Leaving the dépôt in the city, the route passes through the suburbs to the Schuylkill at "Gray's Ferry," which it crosses on a substantial bridge, and thence passes onward via Chester, Wilmington, Delaware, Newport and Elkton, crossing the Susquehanna where it empties into the Chesapeake Bay at Havre-de-Grace: thence 37 miles beyond to Baltimore.

Wilmington, the most important town on this route, is situated between the Brandywine River and Christiana Creek, one mile above their junction, and in the midst of one of the finest agricultural districts in the Middle States. It is built on ground gradually rising to the height of 112 feet above tide-water, and is regularly laid out, with broad streets crossing each other at right angles. Since 1840 both its business and population have much increased: at that time it contained about 8,000 inhabitants, and now the population numbers about 17,000. On the Brandywine River are some of the finest flouring-mills in the United States, to which vessels can come drawing eight feet of water. It contains also ship and steamboat yards, a foundry for the manufacture of patent car-wheels, which are used all over the country, and a number of large manufacturing establishments of various kinds. It is the seat of a Catholic College, and is generally distinguished for its Academies and Boarding-schools. It is connected with New Castle, Dover, and Seaford by railway; and via Downingstown with the Columbia Railroad, from Philadelphia to Columbia.

Havre-de-Grace is in Maryland, at the head of the Chesapeake Bay, on the Susquehanna River, 36 miles north-east of Baltimore. The cars cross the river by a steam ferry, sometimes passing in winter upon the ice as in 1851–2. Havre-de-Grace is quite an old town. It is the southern terminus of the tide water canal.

STEAMBOAT AND RAILROAD ROUTE FROM PHILADELPHIA TO BALTIMORE.—A steamboat leaves from the foot of Dock street every afternoon, and proceeds down the Delaware to *New Castle*, 36 miles; thence by the cars over the New Castle and Frenchtown Railroad to *Frenchtown*, 17 miles. At the latter place, again take the steamboat, and pass down Elk River and Chesapeake Bay, and up the Patapsco River to Baltimore, 60 miles.

PENNSYLVANIA.

PENNSYLVANIA is, in point of population, the second State in the Union, and in all respects one of the most important and interesting. A very singular fact in her history—singular because it has no parallel in the annals of any other member of the American Confederacy—is, that her territory was settled without war or bloodshed. The doctrines of peace and good-will, taught by William Penn and his quiet-loving associates, when they pitched their tents upon the sunny banks of the Delaware, long served, happily, as a charm over the savage natures of their Indian neighbors.

We find no record of battle and siege in the story of this State, from the time of the first settlement at Philadelphia, in 1682, until the date of the French and Indian war in 1755. During this year the famous defeat of Braddock, in which Washington, then in his early youth, distinguished himself, occurred at Pittsburg. In 1763, the massacre of the Conestoga Indians took place in Lancaster County. In 1767, the southern boundary of the State, which has since become famous as Mason and Dixon's line, was made. This line is the proverbial division between the Northern, or Free, and the Southern, or Slave-holding States.

Pennsylvania is memorable in the annals of the American Revolution, in which she played a conspicuous part. Upon her soil occurred the important battles of Brandywine and Germantown (1777).

The traveller will seek here, also, for the scenes of those celebrated events, the massacres of Wyoming and Paoli. Valley Forge was the chief head-quarters of General Washington, and is made yet more interesting by the memory of the sufferings there of the patriot army during its winter encampment in 1777 and 1778. Philadelphia was the national capital until 1789—a period of nearly ten years—and here the earliest American Congresses assembled. The memorable revolt called the Whiskey Insurrection, happened in Pennsylvania in 1794. This disaffection was bloodless and without sequence, as all disloyalty must ever be in the Keystone State.

Among the great men whom Pennsylvania has given to the Republic, we may cite the honored names of Franklin (though born in Boston), Robert Morris, Fulton, Rush, and Rittenhouse. Mr. Buchanan, the present President, is a citizen of this State.

The landscape of Pennsylvania is extremely diversified and beautiful. One-fourth of her great area of 46,000 square miles is occupied by mountain ranges, sometimes reaching an elevation of 2,000 feet. These hills, links of the great Alleghany chain, run generally from north-east to south-west, through the eastern, central, and southern portions of the State. The spur of this hill-range is called South Mountain, where it rises on the Delaware, below Easton. Next, as we go westward, come the Kittatinny, or Blue Mountains, and the Broad Mountain, south of the North Branch of the Susquehanna. Across this river is the Tuscarora. South of the Juniata are the Sideling Hills, and, lastly, come the Alleghanies, dividing the Atlantic slope from the great Mississippi Valley region. West of the Alleghanies, the only hill-ranges in the State are the minor ones called the Laurel and the Chestnut ridges. This belt of mountains extends over a breadth of 200 miles, enclosing numberless fertile valleys, many charming waters, and the greatest coal fields and iron deposits in the Union.

RIVERS.—Pennsylvania cannot boast the marvellous lake-scenery of the Empire State; indeed, she has no lakes, if we except the great Erie waters which wash the shore of the north-west corner of the State. For this want, however, the charms of her many picturesque rivers well atone.

The Susquehanna, the largest river of Pennsylvania, and one of the most beautiful in America, crosses the entire breadth of the State, flowing 400 miles in many a winding bout, through mountain gorges, rocky cliffs, and broad cultivated meadows. See Susquehanna River.

The Juniata is the chief affluent of the Susquehanna. It comes in from the acclivities of the Alleghanies in the west, through a mountain and valley country of great natural delight. See Juniata River.

The Delaware flows 300 miles from its sources in the Catskill Mountains to the Delaware Bay, forming the boundary between Pennsylvania and New Jersey, and afterwards between New Jersey and Delaware. It is one of the chief features of the varied scenery of the New York and Erie Railway, which follows its banks for 90 miles. (See N. Y. and Erie R. R.) Lower down, its passage through the mountains forms that great natural wonder of the State, the Delaware Water Gap. The rocky cliffs here rise perpendicularly to a height of nearly 1,200 feet. (See Delaware Water Gap.) The navigation of the Delaware is interrupted at Trenton, N. J., by falls and rapids. Philadelphia is on this river, about 40 miles above its entrance into Delaware Bay. The river was named in honor of Lord De La Ware, who visited the bay in 1610.

The Lehigh is a rapid and most picturesque stream. Its course is from the mountain coal districts, through the famous passage of the Lehigh Water Gap below Mauch Chunk, to the Delaware at Easton. Its length is about 90 miles.

The Schuylkill flows 120 miles from the coal regions north, and enters the Delaware five miles below Philadelphia. We shall review it as we call at the towns and places of interest upon its banks.

The Alleghany and the Monongahela Rivers—one 300 and the other 200 miles in length—unite at Pittsburg and form the Ohio. The Youghiogheny is a tributary of the Monongahela.

PHILADELPHIA AND VICINITY.*

This great city is, in extent and population, the second in the Union. Its people number about half a million—as many as any of the capitals of Europe (London and Paris only excepted) can show. It was settled in 1682 by a colony of English Quakers, under the guidance of William Penn. The soubriquet of the City of Brotherly Love, which it now bears, was given to it by Penn himself. No striking events mark its history down to the days of the Revolution, and its part in that great drama was more peaceful than warlike. The first Congress assembled here, and subsequent Congresses, during the continuance of the war. The Declaration of Independence was signed and issued here, July 4th, 1776. The Convention which formed the Constitution of the Republic assembled here, May, 1787. Here resided the first President of the United States, and here, too, Congress continued to meet until about 1797. The city was in possession of the British troops from Sept., 1777, to June 11, 1778, a result of the unfortunate battles of Brandywine and Germantown.

Philadelphia lies between the Delaware and Schuylkill rivers, six miles above their junction, and nearly 100 miles (by the Delaware River and Bay) from the Atlantic. The site of the city is so low and level that it does not make a very impressive appearance from any approach. But the elegance and symmetry and neatness of its streets—the high cultivation of all its rural corners, and the picturesque character of the higher suburban land to the northward, fully compensate for this want. The most thronged portion of the city is near the apex of an angle formed by the approach of the two rivers, between which it is built. Streets extend from river to river, and are crossed by other streets at right angles.

PUBLIC   SQUARES. — *Washington*

* See "New Jersey" for routes to New York, and "Delaware" for routes to Baltimore.

SCHUYLKILL RIVER

WEST PHILADELPHIA

Bridgewater Suspens[.]

Thirty First

Callowhill

William

Columbia

Water

Bank

Carbon

Willow

Beach

S. Twenty Third

South Twenty Second

N. Twenty

Aspen Alley

S. Twenty First

N.

N. N.

S. Twentieth

Asylum

N.

N. Twentieth

S. Nineteenth

Trus.[t] for the Bli[nd]

Logan Square

N. Nineteenth

Rittenhouse Square

N. Eighteenth

S. Eighteenth

Wills Hospital

Parkview

S. Seventeenth

Orphans Asylum

Seventeenth

Morris

Lane

Widows Asylum

Vine

Sixteenth

S. Sixteenth

Centre

Lane

S. Fifteenth

N. Fifteenth

Hamilton

Spruce

South

Front

Chestnut

Gas Works

Market

---

ur[.]rental

19

18

ospberry Al.

Univer Penns[.]a

20

arleberry Al.

17

Musical

Post Office

Penn B[.]k

Union

Penn

K

Dock

Front

Almond

water

Penn

Blackhorse

James A[.]

Coombs

Eckets Al.

Prebbers Al.

Pools Al.

Keys Al.

N.

DELAWARE RIVER

Scale of Feet

00

PHILADELPHIA.

Scenes and Places in Philadelphia.

The State House, or Independence Hall, Philadelphia.

Square, a little south-west of the State House, is finely ornamented with trees and gravelled walks, is surrounded by a handsome iron railing, with four principal entrances, and is kept in excellent order. *Independence Square*, in the rear of the State House, is enclosed by a solid brick wall, rising three or four feet above the adjacent streets, surmounted by an iron railing. The entire area is laid off in walks and grass-plots, shaded with majestic trees. It was within this enclosure that the Declaration of Independence was first promulgated, and at the present day it is frequently used as a place of meeting for political and other purposes. *Franklin Square*, between Race and Vine, and Sixth and Franklin streets, is an attractive promenade, with a fountain in its centre, surrounded by a marble basin; it is embellished with a great variety of trees. *Penn Square* is at the intersection of Broad and Market streets, now divided into four parts by cutting Market and Broad streets through it; *Logan Square* is between Race and Vine streets; and *Rittenhouse Square*, between Walnut and Locust streets.

PUBLIC BUILDINGS.—The State House fronts on Chestnut street, and including the wings, which are of modern construction, occupies the entire block,

extending from Fifth to Sixth streets. In a room in this building, on the 4th of July, 1776, the Declaration of Independence was adopted by Congress, and publicly proclaimed from the steps on the same day. The room presents now the same appearance it did on that eventful day, in furniture and interior decorations. This chamber is situated on the first floor, at the eastern end of the original building, and can be seen by visitors on application to the person in charge of the State House. In the Hall of Independence is a wooden statue of Washington, and some pictures. Visitors may overlook the city and its surroundings admirably from the cupola of this building.

The *Girard College* is situated on the Ridge Road, in a north-west direction from the city proper, about two miles from the State House. It was founded by the late Stephen Girard, a native of France, who died in 1831, and bequeathed $2,000,000, for the purpose of erecting suitable buildings for the education of orphans.

The commanding site of the edifice occupies an area of about 45 acres, left for the purpose by the founder of the institution. The central, or college building, is 218 feet long, 160 broad, and 97 high, and is a very noble marble structure of the Corinthian order. The other buildings, six in number, surround the main edifice.

The *Merchants' Exchange*, situated between Dock, Walnut, and Third sts., is of white marble. It is a beautiful structure, and of its kind, one of the finest in the country.

The *United States Mint* is in Chesnut street, below Broad street, and fronts on the former street 122 feet. It is built of white marble, in the style of a Grecian Ionic temple, and comprises several distinct apartments. Coining is among the most interesting and attractive of processes, to those who have never witnessed such operations. Visitors are admitted during the morning of each day, until one o'clock, on application to the proper officers.

The *Arcade* is in Chestnut, between Sixth and Seventh streets, and extends through to Carpenter street, 150 feet, fronting 100 on Chesnut street. On the ground floor are two avenues, extending the entire depth of the building, with rows of stores fronting on each. The second floor of this building was also at one time used, like the lower part, for stores, and yet above was Peale's Museum, a famous resort in its day. The glory both of the Arcade and of the Museum has, however, long since departed.

The *Custom House*, formerly the United States Bank, is located in Chesnut street, between Fourth and Fifth streets. It is a chaste specimen of the Doric order of architecture, after the Parthenon at Athens, with the omission of the colonnades at the sides. It was commenced in 1819, and completed in about five years, at a cost of half a million of dollars.

The *United States Navy Yard* is located in Front street, below Prime, and contains within its limits about 12 acres. It is enclosed on three sides by a high and substantial brick wall; the east side fronts on and is open to the Delaware River. Its entrance is in Front street. The Yard contains every preparation necessary for building vessels of war, and has marine barracks, with quarters for the officers.

Many of the bank edifices of Philadelphia are very elegant, and imposing, built of marble and other rich material.

The **Churches** of the city are about 300 in number, of all denominations, and new ones are continually making their appearance.

The Catholic *Church of St. Peter and St. Paul*, on Logan Square, is built of red stone, in the Roman style. It is crowned with a dome 210 feet high.

The *Church of St. Mark's* (Episcopal), is a beautiful edifice of light-red sand-stone, with a tower and steeple of admirable grace.

Christ's Church, with its soaring spire, is a very interesting object in its ancient and quaint aspect.

The Church of Calvary (Presbyterian), and the Baptist Church in Broad and Arch streets, are also of sand-stone, with imposing towers and spires. We may also mention among the churches of the greatest architectural interest: St. Stephen's (Episcopal), the Catholic Church of the Assumption, St. Jude's, the Presbyterian Churches, upon Arch and Eighteenth streets, and upon Arch and Tenth streets; the Church of the Nativity, the Baptist Churches on Chesnut and Fifth streets. In the towers of St. Peter's, St. Stephen's, and of Christ Church, there are chimes of bells.

The American Baptist Publication Society is located in Arch street; the Presbyterian Board of Publication is at 265 Chesnut street. Besides these religious associations, there are the American Sunday School Union, the Pennsylvania and the Philadelphia Bible Societies, and the Female and the Friends' Bible Societies, with numerous others.

BENEVOLENT INSTITUTIONS.—The county *Almshouse*, situated on the west side of the Schuylkill, opposite South street, is an immense structure, consisting of four main buildings, covering and enclosing about 10 acres of ground, and fronting on the Schuylkill River. The site is much elevated above the bank of the river, and commands a fine view of the city and surrounding country.—The *Pennsylvania Hospital*, in Pine street, between Eighth and Ninth streets, is an admirable institution. It contains an anatomical museum, and a library of more than 8,000 volumes. In the rear of the lot fronting on Spruce street, is a small building which contains West's celebrated picture of Christ Healing the Sick, presented to this institution by its author.—The *United States Marine Hospital or Naval Asylum* has a handsome situation on the east bank of the Schuylkill below South street. It is for the use of invalid seamen, and officers disabled in the service.—*The Pennsylvania Institution for the Deaf and Dumb* is situated on the corner of Broad and Pine streets, having extensive buildings adapted for the purposes of the establishment.—The *Pennsylvania Institution for the Instruction of the Blind* is situated in Race street, corner of Twenty-first street.

ART SOCIETIES. The *Pennsylvania Academy of Fine Arts*, an old and most important institution, has a fine building, with a noble suite of galleries upon Chestnut street, between Tenth and Eleventh streets. It possesses a very valuable and permanent collection of pictures, and makes an annual exhibition of new works. Among its old pictures, are West's Death on the Pale Horse, and Alston's Dead Man Restored. No citizen or stranger should neglect to visit these galleries.

LITERARY AND SCIENTIFIC INSTITUTIONS. The *American Philosophical Society* was founded in 1743, principally through the exertions of Dr. Franklin; its hall is situated in South Fifth street, below Chestnut, and in the rear of the State House. In addition to its library of 15,000 volumes of valuable works, the society has a fine collection of minerals and fossils, ancient relics, and other interesting objects. Strangers are admitted to the hall on application to the librarian.—The *Philadelphia Library* is situated in Fifth street, below Chestnut, on the north corner of Liberty street. It was founded in 1731 by the influence of Dr. Franklin. This institution, together with the Loganian, which occupies the same building, possesses about

65,000 volumes.—The *Athenæum*, in Sixth below Walnut street, contains the periodical journals of the day, and a library consisting of several thousand volumes. The rooms are open every day and evening (Sundays excepted) throughout the year. Strangers are admitted gratuitously for one month, on introduction by a member.—The *Franklin Institute* was incorporated in 1824; it is situated in Seventh street, below Market. Its members are very numerous, composed of manufacturers, artists, mechanics, and persons friendly to the mechanic arts. The annual exhibitions of this institute never fail to attract a large number of visitors. It has a library of about 6,000 volumes, and an extensive reading room, where most of the periodicals of the day may be found. Strangers are admitted to the rooms on application to the actuary.—The *Academy of Natural Sciences*, incorporated in 1817, has a well-selected library of about 14,000 volumes, besides an extensive collection of objects in natural history. Its splendid hall is in Broad street, between Chestnut and Walnut. It is open to visitors every Saturday afternoon. —The *Mercantile Library*, situated on the corner of Fifth and Library streets, was founded in 1822, for the purpose of diffusing mercantile knowledge.— The *Apprentices' Library*, corner of Fifth and Arch streets, consists of about 14,000 volumes, and is open to the youth of both sexes.—The *Historical Society of Pennsylvania*, in Fifth street, below Chestnut, was founded for the purpose of diffusing a knowledge of local history, especially in relation to the State of Pennsylvania. It has caused to be published a large amount of information on subjects connected with the early history of the State, and is now actively engaged in similar pursuits.—The *Friends' Library* in Race street, below Fifth, has about 3,000 volumes, which are loaned, free of charge, to persons who come suitably recommended.—There are several excellent libraries in the Districts of Philadelphia, which are conducted on the most liberal principles.

MEDICAL INSTITUTIONS. The *University of Pennsylvania*, which is an admirable institution, is situated on the west side of Ninth street, between Market and Chestnut. It was founded in 1791, by the union of the old University and College of Philadelphia.—*Jefferson Medical College* is situated in Tenth street, below Chestnut; it was originally connected with the college at Canonsburg, but it is now an independent institution. The number of pupils averages about 300 annually. The anatomical museum of this institution is open to visitors.— *Pennsylvania Medical College*, in Filbert street, above Eleventh, is a flourishing institution of recent origin; the first lectures having been delivered in the winter of 1839-40.—The *College of Physicians* is an old institution, having existed before the Revolution. It is one of the principal sources from which proceeds the Pharmacopœia of the United States.—The *Philadelphia College of Pharmacy*, in Zane street, above Seventh, was the first regularly organized institution of its kind in the country. Its objects are to impart appropriate instruction, to examine drugs, and to cultivate a taste for the sciences. The medical schools of Philadelphia are famous, all the Union through, and the students who flock to them every winter may be numbered by thousands.

PRISONS. The *Eastern Penitentiary*, in the north-west part of the city, is situated on Coates street, cor. of twenty-first street, and south of Girard College. It covers about 10 acres of ground, is surrounded by a wall 30 feet high, and in architecture resembles a baronial castle of the middle ages. It is constructed on the principle of strictly solitary confinement in separate cells, and is admirably calculated for the security, the health, and, so far as consistent with its objects, the comfort of its occupants.— The *County Prison*, situated on Passyunk Road, below Federal street, is a spacious Gothic building, presenting an imposing appearance. It is appropriated to the confinement of persons awaiting trial, or those who are sentenced for short periods. The *Debtor's Prison*, adjoining the above on the north, is constructed of red sandstone, in a style of massive Egyptian architecture.—The *House of Refuge* is situated in Parish street, between Twenty-third and Twenty-fourth streets, and at Bush Hill is the *House of Correction*.

CEMETERIES. The beautiful cemetery of *Laurel Hill* is situated on the Ridge road, three and a half miles north-west of the city, and on the east bank of the Schuylkill, which is elevated about 90 feet above the river. It contains about 20 acres, the surface of which is undulating, prettily diversified by hill and dale, and adorned with a number of beautiful trees. The irregularity of the ground, together with the foliage, shrubs, and fragrant flowers, which here abound—the finely-sculptured and appropriate monuments—with an extensive and diversified view, make the whole scene highly impressive. On entering the gate, the first object that presents itself to the gaze of the visitor is an excellent piece of statuary, representing Sir Walter Scott conversing with Old Mortality, executed in sandstone by the celebrated Thom. The chapel, which is situated on high ground to the right of the entrance, is a beautiful Gothic building, illuminated by an immense window of stained glass. *Monument Cemetery*, another beautiful enclosure, is situated on Broad street, in the vicinity of Turner's Lane, in the north part of Philadelphia, and about three miles from the State House. It was opened in 1838, and now contains many handsome tombs.—*Ronaldson's Cemetery*, in Shippen street, between Ninth and Tenth, occupying an entire square, and surrounded by an iron railing, is very beautiful. It formerly belonged to Mr. James Ronaldson, from whom it takes its name, who divided it into lots, and disposed of it for its present purposes. The *Woodland's Cemetery*, beyond the Schuylkill, in West Philadelphia, is a new burial ground, which promises in due course of time to rival even *Laurel Hill* in landscape and monumental attractions.

PLACES OF AMUSEMENT. The *Acad-*

emy *of Music or Opera House*, on Broad and Locust streets, is a grand establishment, with a front of 140 feet, and a flank of 238. The first story is of brown stone, and the rest of pressed brick with brown-stone dressings. The Auditorium will seat 3000 persons. The *Walnut Street Theatre* is at the corner of Walnut and Ninth streets. *Arch Street Theatre* is in Arch street, above Sixth. The *Musical Fund Hall* is in Locust street, between Eighth and Ninth streets. The *City Museum*, Callowhill, below Fifth ; *Welch's National Circus*, Walnut street, above Eighth ; *Concert Hall*, Chestnut, below Thirteenth ; *National Hall*, Market street, below Thirteenth ; *Sansom Street Hall*, Sansom, above Sixth ; the *Assembly Buildings*, Chestnut and Tenth streets.

Hotels. Philadelphia is abundantly supplied with excellent hotels of all grades. Among the largest, most sumptuous, and most fashionable are, first, the *Continental* (opened 1860), Chestnut, between 8th and 9th streets, in the heart of the city ; the *Girard House*, directly opposite the Continental ; the *La Pierre*, on the west side of Broad, near Chestnut street ; *Jones Hotel*, Chestnut street ; the *Washington Hotel*, in the same vicinage ; the *St. Lawrence*, the *American*, the *United States*, the *Franklin*, and many others.

The *Markets* of Philadelphia are worthy of especial notice, in their great extent and admirable appointment.

Cars and Stages to all parts of the city and suburbs are easily to be found.

THE VICINITY OF PHILADELPHIA. Laurel Hill, and other cemeteries, and the Girard College we have already mentioned.

Camden is upon the opposite side of the Delaware, in New Jersey. It is the terminus of the Camden Railway Route from New York.

The Fairmount Water Works, which supply the city bountifully, are on the east bank of the Schuylkill, about two miles in a northwest direction from the city, occupying an area of 30 acres, a large part of which consists of the "mount," an eminence 100 feet above

tide-water in the river below, and about 60 feet above the most elevated ground in the city. The top is divided into four reservoirs, capable of containing 22,000,000 gallons, one of which is divided into three sections for the purpose of filtration. The whole is surrounded by a beautiful gravel-walk, from which may be had a fine view of the city. The reservoirs contain an area of over six acres ; they are 12 feet deep, lined with stone and paved with brick, laid in a bed of clay, in strong lime cement, and made water-tight. The power necessary for forcing the water into the reservoirs is obtained by throwing a dam across the Schuylkill ; and by means of wheels moved by the water, which work forcing pumps, the water of the river is raised to the reservoirs on the top of the "mount." The dam is 1,600 feet long, and the race upwards of 400 feet long and 90 wide, cut in solid rock. The mill-house is of stone, 238 feet long, and 56 wide, and capable of containing eight wheels, and each pump will raise about 1,250,000 gallons in 24 hours.—The Spring Garden Waterworks are situated on the Schuylkill, a short distance above Fairmount.

The Falls of the Schuylkill are about four miles above the city, on the river of that name. Since the erection of the dam at Fairmount, the falls have entirely disappeared. From the city to the falls, however, is a very pleasant drive ; and they might be reached in a return visit to the Wissahickon.

The Schuylkill Viaduct, three miles northwest from the city, is 980 feet in length, and crossed by the Columbia Railroad. It leads to the foot of an inclined plane, 2,800 feet long, with an ascent of 187 feet. The plane is ascended by means of a stationary engine at the top, which conveys the cars from one end of the plane to the other. It is a pleasant and cheap excursion.

Wissahickon Creek, a stream remarkable for its romantic and beautiful scenery, falls into the Schuylkill about six miles above the city. It has a regular succession of cascades, which in the aggregate amount to about 700 feet.

The Vicinity of Philadelphia.

Fairmount Water Works, Philadelphia.

Its banks, for the most part, are elevated and precipitous, covered with a dense forest, and diversified by moss-covered rocks of every variety. The banks of the beautiful Wissahickon afford one of the most delightful rides in the vicinity of Philadelphia, and are a great resort for the citizens, picnic parties, and Sunday schools.

Manayunk, eight miles from the city, has become a large manufacturing place. It is indebted for its existence to the water created by the improvement of the Schuylkill, which serves the double purpose of rendering the stream navigable, and of supplying hydraulic power to the numerous factories of the village.

Germantown, six miles north of Philadelphia, consists of one street only, compactly built, and extending for about four miles, in a direction from south-east to north-west. A railroad and numerous stages afford a constant communication between this place and the city, of which it is a suburb. Cars leave the depot in Philadelphia, corner of Ninth and Green streets, 12 times daily. Fare 15 cents.

Kaighn's Point, a short distance below Camden, *Greenwich Point,* three miles below the city, and *Gloucester Point,* directly opposite, are favorite places of resort during the summer season. Steamboats run many times daily from Philadelphia. Fare to the former place 5 cents—to the latter, 6¼ cents.

Cape May is a famous watering place at the extreme southern point of New Jersey, just where the floods of the Delaware are lost in the greater floods of the Atlantic. The beach here is of excellent quality for bathing or riding. The hotels, chief among which is the "Congress," are numerous and well appointed. In the season, the little village of the Cape is thronged with thousands of gratified pleasure seekers. They come chiefly from Philadelphia and its vicinage, the access thence being very easy, each day by steamer down the Delaware. The same steamers connect the Cape also with New York, leaving that city every evening. The fare from New York is two dollars; from Philadelphia, one dollar. The trip involves the risk of sea-sickness.

Places of the Reading Railway.

Valley Forge, the memorable head-quarters of General Washington during the winter of 1777, is 23 miles from Philadelphia, on the railway to Reading, Pa. The old head-quarters is still standing near the railroad, from whence it can be seen.

Pottstown, 37 miles from Philadelphia, is prettily situated on the left bank of the Schuylkill. The houses, which are built principally upon one broad street, are surrounded by fine gardens and elegant shade trees. The scenery of the surrounding hills is very fine, especially in the fall of the year, when the foliage is tinged with a variety of rich autumnal tints. The Reading Railroad passes through one of its streets, and crosses the Manatawny on a lattice bridge, 1,071 feet in length.

Reading, 50 miles from Philadelphia by railway, is a pleasant place for a summer home, upon the banks of the Schuylkill river.

Port Clinton is 78 miles from Philadelphia, on the Reading Railroad. It is an agreeable place at the mouth of the Little Schuylkill.

Schuylkill Haven also on the banks of the Schuylkill, in the midst of a very interesting landscape region. A branch road comes in here from the great coal districts. From Philadelphia, 89 miles by Reading Railroad.

Pottsville, the terminus of the Philadelphia and Reading route, is 93 miles from Philadelphia. It is upon the edge of the coal basin, in the gap by which the Schuylkill comes through Sharp's Mountain.

Allentown, 51 miles from Philadelphia, is upon the railroad from Easton, Pa., to Mauch Chunk. It is built upon high ground, near the Lehigh river, at the junction of Jordan and Little Lehigh creeks. The mineral springs here are highly prized by those who have tried the efficacy of their waters. A visit to "Big Rock," 1,000 feet in elevation, a short distance from the village, will amply repay the tourist, by the extent and richness of the scene there spread out before him in every direction.

Bethlehem is upon the Lehigh, near 51 miles from Philadelphia, and 11 from Easton, Pa. May be reached from New York and Philadelphia by railway.

Valley Forge.

Routes to Easton, and thence 12 miles by Lehigh Valley Railroad to Mauch Chunk. It is the principal seat of the United Brethren, or Moravians, in the United States, and was originally settled under Count Zinzendorf, in 1741. The village contains a large stone church of Gothic architecture, 142 feet long and 68 wide, and capable of seating 2,000 persons. From the centre rises a tower, surmounted by an elegant dome.

Nazareth, another pretty Moravian village, is situated 10 miles north from Bethlehem, and 7 miles northwest from Easton.

Mauch Chunk, Pa., is in the midst of the great Pennsylvania coal regions, 43 miles from Easton by railway, and 100 miles from Harrisburg, the State capital. It is upon the Lehigh, in one of its wildest and most romantic passages. Mount Pisgah, a short distance north, rises 1,000 feet along the river. A railway has been constructed, 9 miles, to *Summit Hill*, down which the coal-laden cars come by the force of their own gravity. We are here in the vicinage of the beautiful scenery of the valley of Wyoming and the Susquehanna river, which we shall visit in another chapter.

PHILADELPHIA TO PITTSBURG AND THE WEST.

BY THE PENNSYLVANIA RAILWAY.

This route is one of the great highways from the Atlantic to the Mississippi States. The Pennsylvania Central Road, with some completing links, extends 353 miles, from the city of Philadelphia through the entire length of Pennsylvania to the Ohio river at Pittsburg, connecting there with routes for all parts of the South-west, West, and the North-west. Through trains (15 hours to Pittsburg) run morning, noon, and night. Philadelphia station, southeast corner of Eleventh and Market streets; entrance on Eleventh street.

Lancaster, a city of more than 15,000 inhabitants, is upon the Philadelphia and Columbia Railroad, near the Costenega creek. It was at one time the principal inland town of Pennsylvania,

and was the seat of the State government from 1799 to 1812. In population it now ranks as the fourth in the State. It is pleasantly situated in the centre of a very rich agricultural region, well built, and has many fine edifices, public and private. It is the seat of Marshall College, organized in 1853, in union with the old establishment of Franklin College, which was founded in 1787. Fulton Hall, an edifice for the use of public assemblies, is a noteworthy structure here, as are some of the score of churches. The oldest turnpike road in the United States terminates here, 62 miles from Philadelphia. One of the sources of the prosperity of Lancaster is the navigation of the Costenega, in a series of nine locks and slack water pools, 18 miles in length from the town of Safe Harbor in the Susquehanna, at the mouth of the Costenega. With the help of Tide-Water Canal to Port Deposit, a navigable communication is opened to Baltimore.

Wheatland, the seat of the Hon. James Buchanan, the present President of the United States, is at Lancaster.

Harrisburg, the capital of Pennsylvania, is upon the east bank of the Susquehanna, 106 miles from Philadelphia. From the dome of the State House, a fine view is obtained of the wide and winding river, its beautiful islands, its interminable bridges, and the surrounding ranges of the Kittatinny Mountains. The Cumberland Valley road diverges at Harrisburg for Chambersburg, a flourishing town 52 miles distant, on the south-west, and the Dauphin and Susquehanna Railroad, 59 miles to Auburn, on the Philadelphia and Reading Railroad. The North Central Road is to Baltimore, Md., 85 miles: the Columbia Branch to Columbia. Stop at "Coverly's."

About 14 miles beyond Harrisburg, the route crosses and leaves the Susquehanna river, and thenceforward follows the banks of the Juniata for about 100 miles to the eastern base of the Alleghanies, the canal keeping the road and river company most of the way—of the Juniata part of the route we shall speak directly,—sending the traveller

CONTINUATION OF
ROUTES FROM PHILA
TO
BALTIMORE & WASH.
Also to
POTTSVILLE, HARRISᶜ &
WITH PART OF THE
ROUTES TO PITTSBUR

Drawn & Engraved by W. Williams N.Y.

Unionville

BUTLER

ARMSTRONG

Nicholsburg

KIT ANNING

Middletown

M'Pleasant

INDIANA

Loretto

Birming
ham

Philipsburg

Freeport

Leech
burg

Warren

INDIANA

CAMBRIA

Alexandria

Peters

Altoona

HUNTINGDON

Salzburg

EBENSBURG

HOLLIDAYS
BURG

McCon
nellsburg

BLAIRSVILLE

Johnstown

Martinsburg

PENNSYLVANIA

GREEN'S
BURG

Laughlin T.

Woodberry

Robs?

WASHINGTON
(National)

M'Pleasant

SOMER-
SET

Stoys T.

Shells
burg

Hopewell

BEDFORD

Connellsville

FAYETTE

SOMERSET

ALLEGHENY

BEDFORD

Wereford sburg

McCon
nells T.

UNION

CHESNUT Road

Smithfield

Petersburg

ALLEGHENY

M'Vernon

CUMBERLAND
178 m f! Baltimore

Potomac R.

WASHINGTON and
COCK 124

BATH

Grafton

PRESTON

KINGWOOD

Piedmont

M'Carmee

Frankfort

Spring-
field

MORGAN

MARTINSBURG

BERKELEY

Jamesburg

Wadesley

TAYLOR

BOUR

PHILLIPPA

HAMPSHIRE

Burling
ton

ROMNEY

FREDERICK

WIN-
CHESTER 32 m f! Har
pers Fer.

PHILLIPPA

North Br.

Watsons Spr.

RANDOLPH

HARDY

MOORFIELD

Stephensbg

New T40

WARREN

SHENANDOAH

Sugs
burg

FRONT
ROYAL

BEVERLY

RG

INIA

WOODSTOCK 62

Millwood

FRANKLIN

George T.

M'Pleasant 7X

PENDLETON

Blowing
Cave

Big Spr.

Marb? 80

LURAY

PAGE

MADISON

FAIR
FAX

Rawlys Spr.

HARRISONBURG 98

M'Crawford 105

PAGE

MADISON

ORANGE C.H.

OUTES

MORE

EELING

RINGS.

Woods
boro

Jenning's Gap

Augusta Spr.

Ebbing Spr.

WARM SPRINGS

82 m f! Harpers Fer.

STAUNTON 128

205 from Baltimore

Blowing Cave

Waynesboro

P. Republic

Mt
Sidney

Weirs
Cave

Madison C.H.

ORANGE

Gordonsville 42 m

Fredericksby 132 fr.

ALBEMARLE

University

LOUISA

LOUISA C.H.

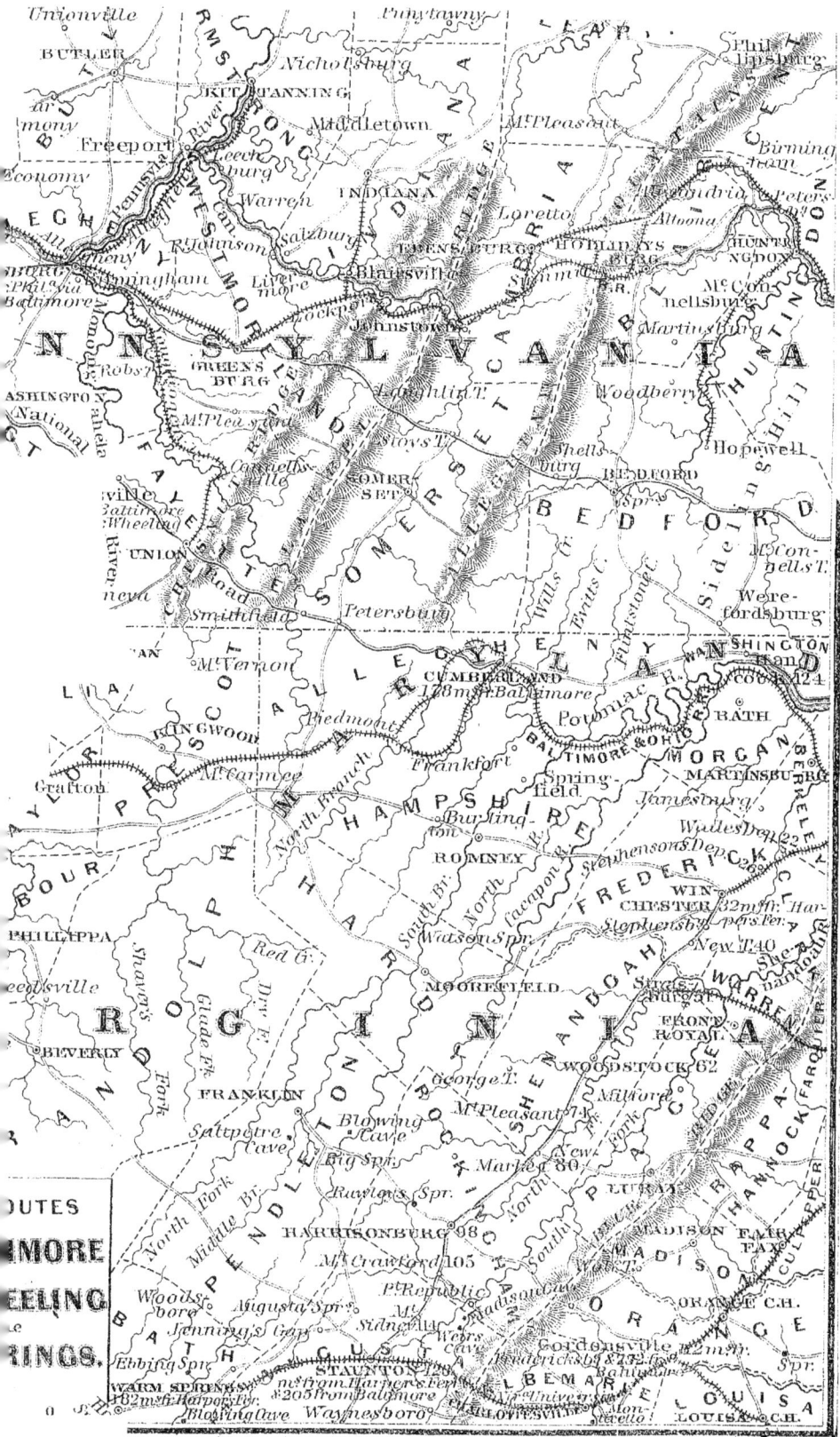

1843, by H. Williams in the Clerks Office of the District Court of the Southern District at New York.

Pittsburg, Pa.

on if he is in haste, to Pittsburg, over the Alleghanies, by the help of the wonderful specimens of the power of the engineer's art, which will interest him on the way : the tunnel, 3,612 feet long, in which he will pass through the Alleghany mountains, 2,200 feet above the sea ; the great inclined planes of the Portage Railroad, and other marvels of art and of nature.

Pittsburg, Pa., is upon the Ohio river, at the confluence of the Alleghany and the Monongahela. It is situated in a district extremely rich in mineral wealth, and the enterprise of the people has been directed to the development of its resources, with an energy and success seldom paralleled. The city of Pittsburg enjoys, from its situation, admirable commercial facilities, and has become the centre of an extensive commerce with the Western States ; while its vicinity to inexhaustible iron and coal mines, has raised it to great distinction as a manufacturing place. The Monongahela House here is one of the finest hotels in the Union.

The city was laid out in 1765, on the site of Fort Du Quesne, subsequently changed to Fort Pitt. It it situated on a triangular point, at the confluence of the Alleghany and Monongahela rivers, which here form the Ohio. Pittsburg is connected with the left bank of the Monongahela by a bridge 1,500 feet long, which was erected at a cost of $102,000 dollars. Four bridges cross the Alleghany river, connecting Pittsburg with Alleghany City.

There are several places in the vicinity of Pittsburg, which, as they may be considered parts of one great manufacturing and commercial city, are entitled to a notice here. *Alleghany City*, opposite to Pittsburg, on the other side of the Alleghany river, is the most important of them. The elegant residences of many persons doing business in Pittsburg, may be seen here, occupying commanding situations. Here is located the *Western Theological Seminary of the Presbyterian Church*, an institution founded by the General Assembly in 1825, and established in this town in 1827. Situated on a lofty, insulated ridge 100 feet above the Alleghany, it affords a magnificent prospect. The *Theological Seminary of the Associated Reformed Church*, established in 1826, and the *Alleghany Theological Institute*, organized in 1840 by the Synod of the Reformed Presbyterian Church, are also located here. The *Western Penitentiary* is an immense building in the ancient Norman style, situated on a plain on the western border of Alleghany

City. It was completed in 1827, at a cost of $183,000. The *United States Arsenal* is located at Lawrenceville, a small but pretty village two and a half miles above Pittsburg, on the left bank of the Alleghany river.

Birmingham is another considerable suburb of Pittsburg, lying about a mile from the centre of the city, on the south side of the Monongahela, and connected with Pittsburg by a bridge 1,500 feet long, and by a ferry. It has important manufactories of glass and iron.

Manchester is two miles below Pittsburg, on the Ohio. The U. S. Marine Hospital is yet below.

It is usual to speak of extensive manufactories as being in Pittsburg, though they are not within the limits of the city proper, but are distributed over a circle of five miles' radius from the court-house on Grant's Hill. This space includes the cities of Pittsburg and Alleghany, the boroughs of Birmingham and Lawrenceville, and a number of towns and villages, the manufacturing establishments in which have their warehouses in Pittsburg, and may consequently be deemed, from the close connection of their general interests and business operations, a part of the city. There are within the above compass about eighty places of religious worship, and a population of not less than 100,000.

The stranger in Pittsburg will derive both pleasure and instruction by a visit to some of its great manufacturing establishments, particularly those of glass and iron. During the summer season Pittsburg is an immense thoroughfare, large numbers of travellers and emigrants passing through it on their way westward. The population of Pittsburg is about 111,000.

The Juniata. This beautiful river, whose course is closely followed so many miles by the Pennsylvania Railroad and Canal, rises in the south central part of the Keystone State, and flowing eastward falls into the Susquehanna about 14 miles above Harrisburg.

The landscape of the Juniata is in

The Juniata.

the highest degree picturesque, and many romantic summer haunts will be found, by and by, among its valleys; though at present very little tarry is made in the region, from its attractions being unknown, and the comforts of the traveller being as yet unprovided for. The mountain background, as we look continually across the river from the cars, is often strikingly bold and beautiful. Our picture is a scene in the upper part of the river, near Water street, a point which the railroad leaves some miles to the south, or left. The Little Juniata, which with the Frankstown branch form the main river, is a stream of wild romantic beauty. The entire length of the Juniata, as well as its branches, is estimated at nearly 150 miles, and its entire course is through a region of mountains, in which iron ore is abundant, and of fertile limestone valleys.

THE COAL REGION.

From Philadelphia.

The *Philadelphia and Reading Railway* extends 93 miles from Philadelphia to Pottsville, in the heart of the *great Coal regions* of the State. It passes through Valley Forge, Reading, Auburn and other places, for which see Index.

The *Catawissa, Williamsport and Erie Railway*, connects Philadelphia with the Erie Railway at Elmira, N. Y., and by other routes from that point with Niagara Falls and all the lines from New York to the great West and Northwest. It leads to the coal beds of Pennsylvania at Catawissa on the Susquehanna, and thence up the west branch of that river to Williamsport. The entire passage of this road is amidst natural scenes of great variety and beauty.

The North Pennsylvania Railroad extends 33 miles, to Doylestown, Pa.

The Belvidere, Delaware and Flemington Railroad extends, via Easton, Pa. (50 miles), to Belvidere, 64 miles.

THE SUSQUEHANNA AND ITS VICINAGE.

We will now look at the chief scenes

and places of interest in Pennsylvania, lying upon and about the great Susquehanna River and its tributaries, and at the railways, canals and other highways of travel which communicate with and intersect that part of the State.

The Susquehanna is the greatest of the rivers of Pennsylvania, traversing as it does its entire breadth from north to south, and in its most interesting and most important regions. It lies about midway between the centre and the eastern boundary of the State, and flows in a zig-zag course, now southeast and now south-west, and so on over and over, following very much the windings of the Delaware, which separates the State from New Jersey. The Pennsylvania Canal accompanies it in all its course, from Wyoming on the north to the Chesapeake Bay on the south. All the great railroads intersect or approach its waters at some point or other, and the richest coal lands of the State lie contiguous to the borders.

The Susquehanna in its main branch rises in Otsego Lake, in the S. E. central part of New York, and pursues a very tortuous but generally south-west course. This main, or North, or East Branch, as it is severally called, when it reaches the central part of Pennsylvania—after a journey of 250 miles—is joined at Northumberland by the West Branch, which comes in 200 miles from the declivities of the Alleghanies. The course of this arm of the river is nearly eastward, and, as with the North Branch, through a country abounding with coal, and other valuable products. It is also followed by a canal, for more than a hundred miles up.

The route of the New York and Erie Railway is upon or near the banks of the Susquehanna in southern New York, and occasionally across the Pennsylvania line, for 50 miles, first touching the river near the Cascade Bridge, nearly 200 miles from New York, passing the cities of Binghamton and Owego, and finally losing sight of it just beyond Barton, some 250 miles from the metropolis. The tourist seeking the picturesque regions of the river from New York, may take the Erie Route, 201 miles to Great Bend, and thence southward by the Delaware, Lackawanna and Western Road, via Scranton, and stage to Wilkesbarre, in the valley of Wyoming. This railway continues on to Lehigh and Easton (Delaware Water Gap) and Elizabethport, back to New York.

The Catawissa, Williamsport and Erie Railway, connects Philadelphia with Catawissa in a beautiful part of the main arm of the Susquehanna below the Wyoming region, and with Williamsport, in the finest part of the West Branch, continuing on through Elmira, N. Y., to the Falls of Niagara. From Philadelphia, via *Port Clinton*, on the Reading Railroad, to Catawissa 145 miles, to Williamsport 197 miles. By this route passengers may go through from Philadelphia to Buffalo in 16 hours, to Niagara Falls in 18 hours, to Detroit in 26 hours, to Chicago in 36 hours, to St. Louis in 48 hours. Day express from Philadelphia breakfasts at Port Clinton and dines at Williamsport.

The Great Pennsylvania Railroad, via Pittsburg to the West, follows the Susquehanna from the vicinity of Harrisburg some 14 miles up to the mouth of the Juniata.

The Northern Central Road from Baltimore, touches the Susquehanna at Harrisburg, 85 miles distant, where it connects with the Pennsylvania Railroad for Pittsburg.

A branch road from Harrisburg follows the river down 28 miles to Columbia.

A pleasant route from Philadelphia or New York to the Valley of Wyoming, is by railway from either city to Easton, near the Delaware Water Gap, thence by the Lehigh Valley Road to the coal regions at Mauch Chunk, and thence to Wilkesbarre.

The entire length of the Susquehanna (or Crooked River) is about 500 miles, and the country which it traverses is of every aspect, in turn, from the gentlest pastoral air to the wildest humors of the stern mountain pass. The region most sought, and deservedly so,

by the tourist in quest of landscape beauties, is that around and below the Valley of Wyoming. From this point down many miles to Northumberland, where the West Branch comes in, the scenery is everywhere strikingly fine at brief intervals; but the best and boldest mountain passes extend from five to ten miles below the southern outlet of Wyoming; around Nanticoke and Shickshinney. This is the region *par excellence* for the study of the artist. Portions also of the West Branch—though not yet very much visited—are remarkably fine.

The Valley of Wyoming and Wilkesbarre.—At Wilkesbarre, in the heart of the Wyoming Valley, there is (near the river) a most excellent hotel. The village is beautifully placed upon a plain 20 feet above the river. Prospect Rock, three miles distant, overlooks the Valley most charmingly.

"Wyoming," says Mr. Minor, in a pleasant history of this vicinity, "though now generally cleared and cultivated, yet to protect the soil from floods, a fringe of trees is left along each bank of the river—the sycamore, the elm, and more especially the black walnut—while here and there, scattered through the fields, a huge shell-bark yields its summer shade to the weary laborers, and its autumn fruit to the black or gray squirrel or the rival plough-boys. Pure streams of water come leaping from the mountains, imparting health and pleasure in their course, all of them abounding with the delicious trout.— Along these brooks and in the swales scattered through the uplands, grow the wild-plum and the butternut; while, wherever the hand of the white man has spared it, the native grape may be gathered in unlimited profusion."

* "Wyoming is a classic and a household name. At our earliest intelligence it takes its place in our hearts as the label of a treasured packet of absorbing history and winning romance. It is the key which unlocks the thrilling

* The Author in Harper's Magazine for October, 1858, vol. vii, p. 615.

The Valley of Wyoming.

recollections of some of the most tragical scenes in our national history, and some of the sweetest imaginations of the poet. Every fancy makes a Mecca of Wyoming, Thus sings Halleck—

When life was in its bud and blossoming,
  And waters gushing from the fountain
    spring
Of pure enthusiast thought, dimm'd my
    young eyes,
As by the poet borne on unseen wing,
  I breathed in fancy 'neath thy cloudless
    skies,
The summer's air, and heard her echoed har-
    monies.'

"The pen of Campbell and the pencil of Turner have taken their loftiest and most unbridled flights in praise of Wyoming, and though they have changed, they have not flattered its beauties.

'Nature hath made thee lovelier than the
    power
Even of Campbell's pen hath pictured.'

"Again, Halleck says of the mythical Gertrude, the fair Spirit of Wyoming, and of the real maidens of the land —

But Gertrude, in her loveliness and bloom,
  Hath many a model here; for woman's eye

In court or cottage, whereso'er her home,
  Hath a heart-spell too holy and too high
To be o'erprais'd, even by her worshipper —
  ♦ Poesy!'"

The terrible Battle of Wyoming— which has been so often the theme of the pencil and the pen, occurred on July 3d, 1778. Few of the ill-fated people escaped. Prisoners were grouped around large stones, and were murdered with the tomahawk, amidst yells and incantations of fiendish triumph. One of these stones of inhuman sacrifice may yet be seen in the Valley. It is called Queen Esther's Rock, and lies near the old river bank, some three miles above Fort Forty. The village of Wilkesbarre was burned at this time, and its inhabitants were killed or taken prisoners, or scattered in the surrounding forests.

The site of Fort Forty is across the river from Wilkesbarre, past the opposite village of Kingston, and nearly west of Troy, five miles and a half distant. At this spot, where the slain were buried, there now stands a monument commemorative of the great disaster. It is an obelisk 62½ feet high, made of granite blocks hewn in the

The Susquehanna.

neighborhood. The names of those who fell and of those who were in the battle and survived, are engraved upon marble tablets set in the base of the monument. This praiseworthy work was done by the exertions of the ladies of Wyoming.

Nanticoke and **West Nanticoke** are little coal villages, at the southern extremity of the Wyoming Valley, where as we have already intimated occur some of the boldest passages of the scenery of the Susquehanna. This point, as others upon the banks of the river below, must be reached by stage, or by the slow and heavily laden canal boats, for railways do not yet traverse the way; and neither are there any better accommodations than those of ordinary village and wayside inns: at least not until we reach Catawissa or Northumberland, where the West Branch comes in. A beautiful view of Wyoming is seen looking northward from the hills on the east side of the river, near Nanticoke; and the scene below, from the banks of the river and the canal, are most varied and delightful. The coal mines of this neighborhood may easily be penetrated, and with ample remuneration for the venture.

Jessup's is a very cosy, lone inn, upon the west shore, two or three miles below Nanticoke, from whence are seen striking pictures of the river and its strong mountain banks both above and below; the hills in all this vicinity are impressively bold and lofty, making the comparatively narrow channel of the river seem yet narrower, and *italicizing* the quiet beauty of the many verdant islands which stud the waters here.

Shickshinney and **Wapwollopen**, yet below, are little places, still in the midst of a rugged hill and valley country. Back of Wapwollopen, on the east shore, is the barren peak of its namesake mountain, and the wild waters of Wapwollopen Creek.

Catawissa is on the line of the railways from Philadelphia for Williamsport, on the West Branch, and thence to Elmira and Niagara. It is connected also by railway with the coal district of Mauch Chunk. The scenery of this vicinity is of great variety and beauty. From the hill-tops—for Catawissa is buried between picturesque hills—remarkable pictures of the winding of the river, and its ever-present companion, the canal, are to be seen—now at the base of grand mural precipices, and, anon, through little verdant intervales.

Northumberland.—The west branch of the Susquehanna unites here with the main, or north arm, and the village, the pleasantest of all the region round, is built upon the point formed by the confluence of the two waters. The quiet, cultivated air of Northumberland, and its excellent hotel, will be very likely to detain the not over-hurried traveller awhile.

Sunbury is a prosperous town across the river. The Sunbury and Erie R.R. connects here with the route from Philadelphia to Williamsport and Elmira, and with the Philadelphia and Sunbury route.

Williamsport is the principal town upon the west branch of the Susquehanna. It is a pleasant place, delightfully situated, and much in vogue as a summer resort. The west branch of the canal passes here; and here, too, the railway routes from Philadelphia and from Niagara Falls meet. The river landscape between Williamsport and Northumberland presents in its long extent many charming passages.

Liverpool is a lively little town upon the Susquehanna and the Pennsylvania Canal, below Northumberland, and 29 miles above Harrisburg.

The *Juniata* River comes into the Susquehanna, 14 miles above Harrisburg. See Juniata in "Pennsylvania R. R. route."

Harrisburg, the capital of Pennsylvania. See "Pennsylvania R. R. route."

Columbia, Pa., the western terminus of the Philadelphia and Columbia R. R., is on the left bank of the Susquehanna, 28 miles below Harrisburg, and 12 west of Lancaster. A part of the town occupies the slope of a hill, which rises gently from the river, and the business part of the town lies along the level bank of the river. The scenery from the hills in the vicinity is highly pleasing. The broad river, studded with numerous islands and rocks, crossed by a long and splendid bridge, and bounded on every side by lofty hills, makes a brilliant display. The junction here of the State railroad from Philadelphia with the main line of the canal, the railroad to York, 12 miles long, and the Tide-water Canal to Maryland, renders Columbia a busy place. The main current of travel, which formerly passed through here, has been diverted by the construction of the Harrisburg and Lancaster Railroad; but the emigrant travel still goes by way of Columbia. A fine bridge crosses the Susquehanna, more than a mile in length.

York, Pa., is ten miles south-west of the Susquehanna, upon the Codorus Creek, 28 miles S. S. E. of Harrisburg, 48 miles from Baltimore, and 92 from Philadelphia. With all these cities, and with yet other points, it is connected by railways. The Baltimore and Susquehanna R. R. unites at York with the York and Cumberland, and with the York and Wrightsville Railroads. The Continental Congress met here in 1777, during the occupation of Philadelphia by the British troops.

Port Deposit is in Maryland, on the east bank of the Susquehanna, at the lowest falls, and five miles from its entrance into the Chesapeake Bay. Fifty millions of feet of lumber are annually floated down the great river, and received at Port Deposit. There are extensive quarries of granite in the neighborhood.

Havre de Grace is at the head of the Chesapeake Bay, at the mouth of the Susquehanna, 36 miles northeast of Baltimore. It is upon the line of the railway from Philadelphia to Baltimore. See that route.

Carlisle, Pa., is a beautiful and interesting town, with a population of 6,000, on the line of the Cumberland Valley R. R., 18 miles below Harrisburg, and 125 miles west of Philadel-

phia. It lies in the limestone valley country, between the Kittatinny and the South Mountains. Dickinson College (Methodist), which is located in Carlisle, is one of the most venerable and esteemed institutions in Pennsylvania. It was founded in 1783. Carlisle is connected by the Cumberland Valley road with Harrisburg, on the one hand, and with Hagerstown, in Maryland, on the other. General Washington's headquarters were here in 1794, at the time of the Whiskey Rebellion. Some years before, Major André was a prisoner of war in Carlisle.

DELAWARE WATER GAP, PA.

The bold passage of the Delaware River, called the Water Gap, is easily and speedily accessible from the cities and vicinage of New York and Philadelphia, and a pleasanter excursion for a day or two cannot be well made. The Delaware River rises on the western declivity of the Catskills, in two streams, which meet at the village of Hancock, a station on the New York and Erie R. R. At Port Jervis (Erie R. R.), after journeying 70 miles, it meets the Kittatinny or Shawangunk Mountain, and next breaks through the bold ridge at the Water Gap. At this great pass the cliffs rise perpendicularly, from 1,000 to 1,200 feet, and the river rushes through the grand gorge in magnificent style. It afterwards crosses the South Mountain, not far below Easton (from which point the Gap is generally approached); next falls over the primitive ledge at Trenton, N. J., grows by-and-by into a large navigable river, skirts the wharves of the city of Philadelphia, and is lost, 100 miles below, in the Delaware Bay. The whole length of this fine river, from the mountains to the bay, is 300 miles.

From New York, take the New Jersey Central road to Easton, Pa., or go from Philadelphia to Easton, in the vicinity of the Water Gap, and thence by other railways. From Great Bend on the Erie Railway, take the Delaware, Lackawanna, and Western Road to the Water Gap.

CONTENTS.

BRITISH AMERICA.

LIST OF MAPS.

LIST OF ILLUSTRATIONS.

INDEX.

PART I.

THE TRAVELLER'S MEMORANDUM.

** The traveller is respectfully solicited to make notes of all errors and omissions which he may discover in this work, and of any new facts of interest,—and to send such memoranda to the Author, care of the Publishers.

CPSIA information can be obtained
at www.ICGtesting.com
Printed in the USA
BVOW06s0740240917
495742BV00019B/474/P